Surfaces, Interfaces, and the Science of Ceramic Joining

Technical Resources

Journal of the American Ceramic Society

www.ceramicjournal.org

With the highest impact factor of any ceramics-specific journal, the *Journal of the American Ceramic Society* is the world's leading source of published research in ceramics and related materials sciences.

Contents include ceramic processing science; electric and dielectic properties; mechanical, thermal and chemical properties; microstructure and phase equilibria; and much more.

Journal of the American Ceramic Society is abstracted/indexed in Chemical Abstracts, Ceramic Abstracts, Cambridge Scientific, ISI's Web of Science, Science Citation Index, Chemistry Citation Index, Materials Science Citation Index, Reaction Citation Index, Current Contents/ Physical, Chemical and Earth Sciences, Current Contents/Engineering, Computing and Technology, plus more.

View abstracts of all content from 1997 through the current issue at no charge at www.ceramicjournal.org. Subscribers receive full-text access to online content.

Published monthly in print and online. Annual subscription runs from January through December. ISSN 0002-7820

International Journal of Applied Ceramic Technology

www.ceramics.org/act

Launched in January 2004, *International Journal of Applied Ceramic Technology* is a must read for engineers, scientists,and companies using or exploring the use of engineered ceramics in product and commercial applications.

Led by an editorial board of experts from industry, government and universities, *International Journal of Applied Ceramic Technology* is a peer-reviewed publication that provides the latest information on fuel cells, nanotechnology, ceramic armor, thermal and environmental barrier coatings, functional materials, ceramic matrix composites, biomaterials, and other cutting-edge topics.

Go to www.ceramics.org/act to see the current issue's table of contents listing state-of-the-art coverage of important topics by internationally recognized leaders.

Published quarterly. Annual subscription runs from January through December.
ISSN 1546-542X

American Ceramic Society Bulletin

www.ceramicbulletin.org

The *American Ceramic Society Bulletin*, is a must-read publication devoted to current and emerging developments in materials, manufacturing processes, instrumentation, equipment, and systems impacting the global ceramics and glass industries.

The *Bulletin* is written primarily for key specifiers of products and services: researchers, engineers, other technical personnel and corporate managers involved in the research, development and manufacture of ceramic and glass products. Membership in The American Ceramic Society includes a subscription to the *Bulletin*, including online access.

Published monthly in print and online, the December issue includes the annual *ceramicSOURCE* company directory and buyer's guide. ISSN 0002-7812

Ceramic Engineering and Science Proceedings (CESP)

www.ceramics.org/cesp

Practical and effective solutions for manufacturing and processing issues are offered by industry experts. CESP includes five issues per year: Glass Problems, Whitewares & Materials, Advanced Ceramics and Composites, Porcelain Enamel. Annual subscription runs from January to December. ISSN 0196-6219

ACerS-NIST Phase Equilibria Diagrams CD-ROM Database Version 3.0

www.ceramics.org/phasecd

The ACerS-NIST Phase Equilibria Diagrams CD-ROM Database Version 3.0 contains more than 19,000 diagrams previously published in 20 phase volumes produced as part of the ACerS-NIST Phase Equilibria Diagrams Program: Volumes I through XIII; Annuals 91, 92 and 93; High Tc Superconductors I & II; Zirconium & Zirconia Systems; and Electronic Ceramics I. The CD-ROM includes full commentaries and interactive capabilities.

Surfaces, Interfaces, and the Science of Ceramic Joining

Ceramic Transactions Volume 158

Proceedings of the 106th Annual Meeting of The American Ceramic Society, Indianapolis, Indiana, USA (2004)

Editors
K. Scott Weil
Ivar E. Reimanis
Charles A. Lewinsohn

Published by
The American Ceramic Society
PO Box 6136
Westerville, Ohio 43086-6136
www.ceramics.org

Surfaces, Interfaces, and the Science of Ceramic Joining

For information on ordering titles published by The American Ceramic Society, or to request a publications catalog, please call 614-794-5890, or visit our website at www.ceramics.org

ISBN 1-57498-179-X

Contents

Joining

Preface

This volume contains the proceedings of "Surfaces, Interfaces, and the Science of Ceramic Joining," a symposium held in Indianapolis, IN, April 18–21, 2004 as part of the 106th Annual Meeting of The American Ceramic Society. With over 50 presentations and posters, the symposium was the successful outgrowth of prior symposia on surface science, interfacial analysis, and ceramic joining. In keeping with our objective to offer a forum for those interested in discussing the fundamental aspects of ceramic surface and interfacial phenomenon and their relationship to the nature of bonding/joining in ceramic materials, a wide range of subject matter was covered during the three days of presentations—from ceramic surface characterization and molecular dynamic modeling to interfacial phenomenon, such as boundary layer transitions between metal/ceramic interfaces in cermet composites and observations on intergranular phase transformations, as well as topics of particular significance to ceramic joining, including wetting, adhesion, and interfacial mechanics.

The breadth of the symposium is well represented in this proceedings volume, which includes papers on: the development of photocatalytic titania coatings, the mechanics of functionally graded ceramic-to-metal joints, new techniques for measuring coating adhesion and ceramic joint strength, characterization of surface wetting as a function of substrate and wetting liquid composition, and the development of chiral surfaces as templates for catalytic thin film growth. We would like to thank all of the participants in the symposium and especially those who contributed to this volume. Many thanks are also due to the staff at The American Ceramic Society for their assistance in handling numerous details before, during, and after the meeting and for helping to produce this proceedings.

K. Scott Weil
Ivar E. Reimanis
Charles A. Lewinsohn

Surface and Interfacial Phenomena

THE ROLE OF INTERFACIAL PHENOMENA IN WETTING-BONDING RELATIONSHIP IN Al/CERAMIC COUPLES

Natalia Sobczak
Foundry Research Institute
73 Zakopianska St
30-418 Krakow, POLAND

Rajiv Asthana
University of Wisconsin-Stout
326 Fryklund Hall
Menomonie, WI 54751, U.S.A.

ABSTRACT

The wetting-interface strength relationship in high-temperature ceramic/metal couples must be interpreted in light of nano- and micro-scale structure of the interface. New experimental results on the effect of liquid-phase joining parameters on the wetting-structure-strength response of Al/Al_2O_3 and Al/TiO_2 couples are discussed. The influence of time, temperature, alloying, ceramic additives, and metal films on Al_2O_3 is examined, and it is observed that non-equilibrium phenomena (segregation, sedimentation, dissolution, and defects) markedly influence the interface behavior. It is argued that data on the classical liquid-state joining science parameters (contact angle and work of adhesion) must be coupled with the structural information to develop a scientific understanding of the joining process.

INTRODUCTION

The wetting-bonding relationships in ceramic/metal couples become increasingly complex at elevated temperatures due to the extreme sensitivity of the nano- and micro-scale structure of the interface to the joining process variables, and a host of ubiquitous imperfections that reside at the interface. Interpretations of wetting-bonding relationship based solely on the classical surface thermodynamic parameters, without consideration of the microstructural, compositional and morphological features of the interface, have led some investigators to conclude that recipes designed to lower the contact angle, θ, or increase the work of adhesion, W_{ad}, might not lead to an increase in the joint strength.

The purpose of this paper is to discuss the effect of liquid-phase joining process parameters on the interface response with a view to understanding the wetting-bonding-interface structure relationship in two technologically important couples: Al/Al_2O_3 and Al/TiO_2. In particular, the role of high-temperature wettability and reactivity in the evolution of the interface structure, and their effect on joint strength will be discussed. The influence of contact time, temperature, alloying additions to Al, ceramic additives in Al_2O_3, and thin Ti and Sn films on Al_2O_3 will be examined. The role of universally-present residual oxygen and non-equilibrium phenomena (e.g., phase segregation, sedimentation, dissolution, and defects) in the evolution of the interface and joint strength will be highlighted. The paper will conclude with the proposition that the classical liquid-state joining science parameters might be inadequate to assess the effectiveness of a joining technology in systems where the interface undergoes substantial metallurgical transformation during joining. For such systems, empirical measures of adhesion strength should yield criteria to design an optimum ceramic/metal joint.

EXPERIMENTAL

Sessile-Drop Test: The contact angles were measured using the sessile-drop method described in Ref. 1. The test is carried out in a dynamic vacuum of ~0.2 mPa for different contact times and at different temperatures. Three sample heating procedures were employed: 1) *fast contact heating*

(FCH) (~40 K/min) was achieved by introducing the couple into the furnace previously heated to the test temperature; 2) *slow contact heating* (SCH) (~10 K/min) was achieved when the couple was first placed in the furnace and then heated to the test temperature; and 3) *capillary purification* (CP) in which the substrate and the metal were heated separately under vacuum, the metal in a graphite syringe. At the test temperature, a droplet of the metal is mechanically squeezed out of the graphite syringe and brought in contact with the preheated ceramic. Separate heating of the metal and the substrate in CP eliminates chemical interactions that would occur during contact heating to the test temperature, while the extrusion of the liquid out of the syringe forms droplets free of oxide film, thus establishing true substrate/metal physical contact. To reduce the thermal stress during heating, the ceramic/metal couple can be heated slowly, and at the conclusion of the test, cooled slowly (~ 10 K/min). With metal-coated substrates, the coating dissolution is minimized by faster heating (40 K/min) and cooling (~20 K/min).

The key test variables in the sessile-drop tests were temperature, time, type of substrate coating (Ti or Sn), the substrate (polycrystalline α-$Al_2O_3^{PC}$ and sapphire single crystals α-$Al_2O_3^{SC}$), and the droplet metal composition (e.g., 99.9999% pure Al, Al–Si, Al–Ti and Al–Sn alloys). The polycrystalline α-Al_2O_3 substrates were sintered at 1923 K from the powder containing less than 0.1% impurities (0.009% CaO, 0.053% SiO_2, 0.0029% MgO, 0.023% Fe_2O_3, and 0.0036% Na_2O). The TiO_2 substrates were hot pressed from powder containing less than 0.1% impurities. All the substrates were polished with diamond paste up to an average roughness of R_a=100-120 nm. Thin coatings (800 nm to <2 μm) were deposited onto one face of selected substrates using physical vapor deposition.

Droplet Push-Off Test: A simple yet elegant approach to relate θ to bond strength is the droplet push-off shear test first employed in early studies[2-5] on Al_2O_3/Me couples (Me = Al, Ni, Ag, or Cu). The push-off test measures the shear stress (applied parallel to the substrate) required to debond solidified sessile-drops from the substrate. A methodological limitation of the droplet push-off shear test is the difficulty in applying a shear stress to thin droplets with $\theta<90°$. An improved push-off test[6] allows shearing of both non-wetting ($\theta>90°$) and wetting ($\theta<90°$) couples because the solidified droplet/ceramic couple is bisected perpendicular to the interface at the mid-plane of the contact circle, and one-half of the bisected droplet is used for bond strength measurement (the other half is either thermally cycled and tested for interface strength, or used for microstructural examination of the joint). For the shear test, a load is applied to the flat end of the bisected couple at a constant rate (1 mm/min), and the load versus displacement data are recorded until failure under shear occurs. By enabling the measurement of θ and τ (shear bond strength) on each individual test specimen, the improved push-off shear test allows characterization of the wetting, bonding, structure, and chemistry of the interface in the *same* test coupon. Recently the test was applied to Al/Al_2O_3[6-8], Al/Si_3N_4[9], Al/AlN[10], Ni/Al_2O_3[11] and Cu/Al_2O_3[12] couples.

RESULTS AND DISCUSSION

Effect of temperature and testing procedure

Much work has been done on measuring the wettability in metal/Al_2O_3 couples. The general conclusion is that alumina is not wetted by Al at the latter's melting point, and the non-wetting-to-wetting transition temperature, T^w, exhibits wide dispersion (1083–1373 K).[5,13-19] T^w depends upon the test technique, furnace atmosphere, substrate roughness, crystal orientation, and chemical purity of the substrate and the metal.

Figure 1(a) shows the θ-time data in SCH Al/Al_2O_3 at 953–1323 K, and in CP Al/Al_2O_3 at 973 K. At 953 K and 1023 K, θ for SCH samples decreases somewhat sluggishly with time, stabilizing at 130° and 128°, respectively, indicating poor wetting. The poor wetting is due to the residual oxygen present in the furnace even under relatively "clean" test conditions, which forms an oxide film on Al droplet that hinders spreading.[13] The θ in CP Al/Al_2O_3 at 973 K shows a marked and rapid (almost immediate upon contact) decrease, and a stable value of ~93° is attained. Such a low θ is not achieved in Al/Al_2O_3 at the low-test temperature of 973 K when contact heating is employed. This indicates that CP causes oxide to rupture, thus yielding a clean, oxide-free solid/liquid interface, and good wetting. Extremely short times are needed to achieve an equilibrium θ in the CP couple even at 973 K; in SCH, this would normally occur only at higher temperatures. At $T>T^w$, θ exhibits a stronger time-dependence, and becomes acute, indicating good wetting (e.g., θ stabilizes at ~80° and ~75° at 1223 K and 1323 K, respectively[8]). The non-wetting-to-wetting transition (θ=90°) occurs at ~1150 K, which agrees with the literature range for T^w (1083-1373 K, depending upon the test conditions).

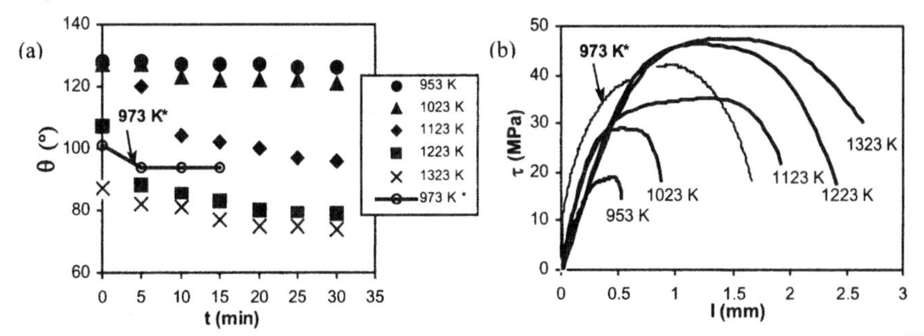

Fig. 1. Effect of temperature of wettability test on (a) wetting and (b) shear behavior of $Al/Al_2O_3^{PC}$ couples: data for SCH from Ref. 6, except data for CP marked by (*) - from Ref. 7.

When oxide is removed from the droplet surface prior to test *via* CP (or when low oxygen partial pressure exists in the furnace), acute values of θ (≤90°) are obtained even at low temperatures ($T<T^w$). On the other hand, at $T>T^w$, the destruction of the oxide film under vacuum aids in lowering of θ. The oxide removal occurs due to 1) the formation of the volatile suboxide, Al_2O by the reaction: $4Al(l) + Al_2O_3(s) \rightarrow 3Al_2O(g)$, and 2) partial dissolution of Al_2O_3 skin in molten Al drop.[13] Simultaneous measurements of θ and oxide thickness in quenched sessile drops (tested under varying oxygen partial pressures, Po_2) have been used[14] to extrapolate the "true" contact angle at zero oxide thickness. Figure 1(b) shows the room-temperature shear stress (τ) versus displacement (l) data in SCH Al/Al_2O_3 couples produced at different wettability test temperatures, and in CP Al/Al_2O_3 at 973 K. The maximum shear stress (τ_{max}) on each curve is a measure of the interfacial shear strength; τ_{max} increases with increasing temperature. The τ_{max} of CP Al/Al_2O_3 is greater than that of SCH Al/Al_2O_3 at $T \leq 1123$ K, and is due to the beneficial effect of mechanical removal of oxide skin from the droplet surface by CP.

The effect of testing procedure on the wetting and shear behaviors of SCH and SCP $Al/Al_2O_3^{PC}$ couples is presented in Fig. 2. For the same temperature of wettability test, the SCH samples show better wetting under higher vacuum (Fig. 2(a)) and the shear behavior (Fig. 2(b)) is consistent with the wettability results. For the same vacuum level and time of interaction, an increase in temperature results in improvement of both wetting and bonding of SCH couples.

However, for different vacuum levels, the time of interaction becomes an important factor, particularly for bonding properties, and this effect is more pronounced for higher temperature of wettability test. Among the SCH samples, the 15 min 1023 K sample produced in a vacuum of 0.4 mPa has the lowest shear strength while its contact angle is slightly lower than that of the 60 min 953 K sample produced in a vacuum of 0.3 mPa, indicating that despite higher temperature more oxide had formed in the former due to higher Po_2 (oxide removal did not occur as $T<T_w$). The CP procedure dramatically lowers the θ, with θ_{eq}~93° at 973 K and θ_{eq} is attained in a very short time unlike the SCH samples. The results of $\theta-\tau$ relationship proves the hypothesis by Laurent *et al.*[18] that the oxide film covering aluminum drop prevents a true contact with ceramic substrate and it is responsible for a lack of low-temperature wettability.

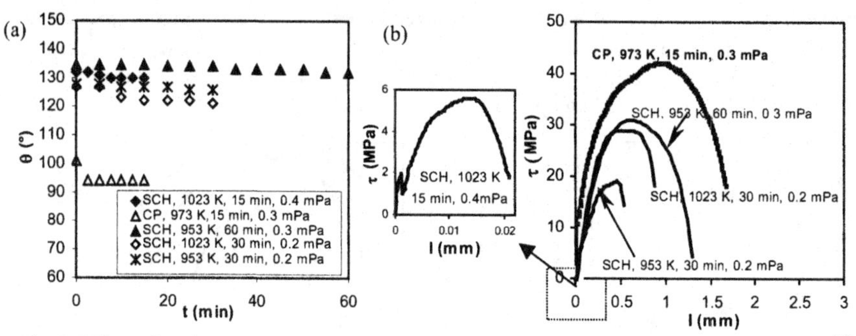

Fig. 2. Effect of testing procedure (SCH, CP) on wetting (a) and shear (b) behavior of $Al/Al_2O_3^{PC}$.

The structural features of the interface in Al/Al_2O_3 sessile-drop couples will now be discussed. Figure 3 shows the scanning electron microscope (SEM) views at three regions of the substrate-side $Al/Al_2O_3^{PC}$ interface: far from TL (Fig. 3(a)), under the drop in the proximity of TL (Fig. 3(b)), and under the droplet near the center (Fig. 3(c)). These figures show that fine Al_2O_3 crystallites are formed at the S/L interface in $Al/Al_2O_3^{PC}$ and $Al/Al_2O_3^{SC}$ (Fig. 3(d)) couples at 1223 K after 30 min contact. Similar observations have been reported by others.[15,17]

The fine Al_2O_3 crystallites on bare substrate far from TL and located in the area surrounding Al droplet in the $Al/Al_2O_3^{PC}$ (Fig. 3(a)) are formed, most probably, due to interaction between evaporated Al and the $Al_2O_3^{PC}$ substrate. Larger (<1 to about 2 μm) crystallites of Al_2O_3 form under the droplet near TL (Fig. 3(b)) on polycrystalline substrate; these crystallites are distributed randomly on Al_2O_3 grains of the substrate rather than preferentially located at pre-existing Al_2O_3 grain boundaries (GB). This suggests that they probably did not form at the GB (although it is possible that crystallites that had nucleated at GB's could have undergone some rearrangement due to liquid flow accompanying solidification contraction during cooling, provided they were not strongly bonded to the substrate). The coarsest (~2-5 μm) Al_2O_3 crystallites are located at the S/L interface near the center of the drop, and these also do not show any evidence of preferential nucleation at pre-existing Al_2O_3 GB's (Fig. 3(c)). The Al_2O_3 crystallites also are noted at S/L interface on single-crystal ($Al_2O_3^{SC}$) substrates as shown in Fig. 3(d). These crystallites have sharp faceted surfaces, and possibly grow epitaxially (i.e., growth of one layer is in a particular crystallographic orientation relationship to the underlying layer). The epitaxial growth could possibly be due to the slow cooling rates employed at the conclusion of sessile-drop tests in our study. It has been suggested[16] that Al_2O_3 crystallites are formed near the drop center by O diffusion along S/L interface through a nanometer size layer. However, as the approximate crystallite size (Fig. 3(d)) is larger than a few nm, it appears that O must diffuse to the growth front (e.g., top face of the crystallite marked with an arrow in Fig. 3(d)) through the crystallite and/or along its surface. As the crystallites appear to be ordered and grow epitaxially, they will likely have a very small defect population (e.g., vacancies) that could assist O diffusion to the growth front. It is possible that mechanisms other than O diffusion along S/L interface could also be playing a role. One possibility is the dissolution of Al_2O_3 of the substrate in Al(l), and reprecipitation of fresh Al_2O_3 from O-saturated Al(l). The dissolution reaction generates Al–O clusters (or an O-rich Al layer) at the interface that decrease σ_{ls} due to adsorption. These clusters could serve as precursor (or seed) to an O-rich interphase, which will lead to the epitaxial growth of fine Al_2O_3 crystallites at the S/L interface especially favorable during slow cooling.

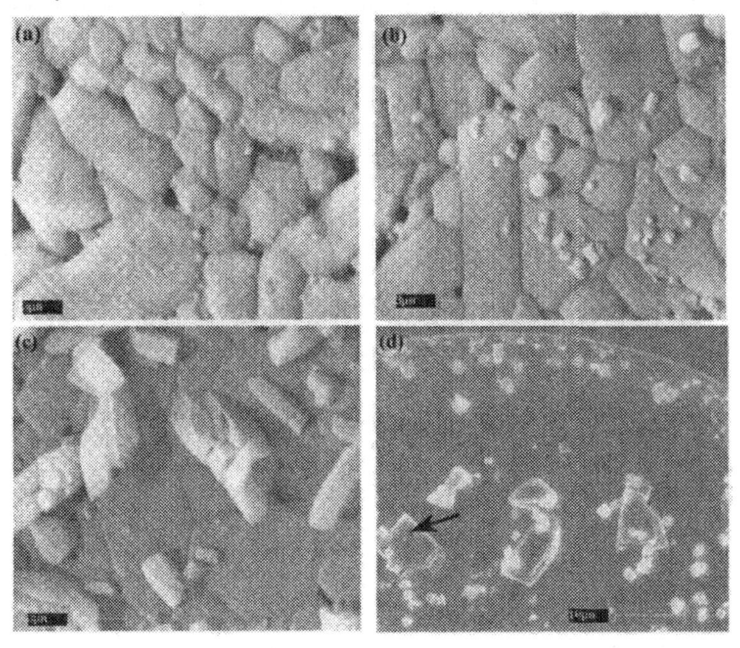

Fig. 3. The droplet/substrate interface structure (1223 K) as revealed by scanning electron microscope; (a) through (c) are for $Al/Al_2O_3^{PC}$ and (d) is for $Al/Al_2O_3^{SC}$: (a) far from TL outside the droplet, (b) under the droplet near TL, (c) at the center under the droplet, (d) under the droplet.

If a dissolution-precipitation process is operative, then the energy released by oxide dissolution could locally raise the temperature at the S/L contact region and the O solubility in the melt. Some studies[20] have actually measured the temperature rise due to chemical dissolution by placing highly sensitive thermocouples exactly "inside" the drop/substrate interface. As the energy released from the dissolution reaction is dissipated, the interface temperature decreases, resulting in the discontinuous precipitation of microcrystalline Al_2O_3. Note that the solubility limit of oxygen in liquid Al could not be precisely determined and it might be much higher than the value given in the literature. Moreover, contrary to literature data, in our study we used Al of spectral purity and we expect even higher amount of alumina to dissolve.

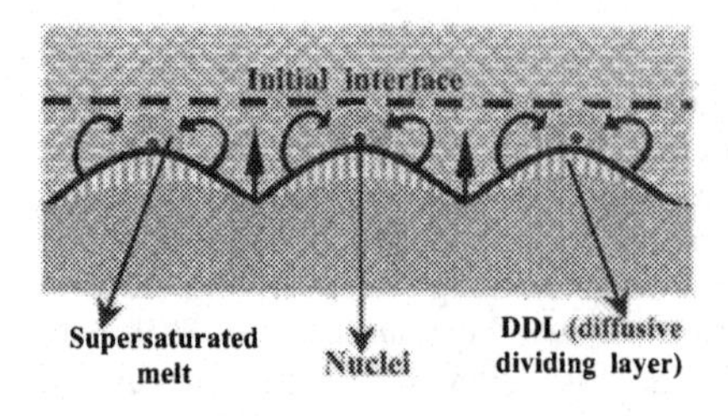

Fig. 4. Schematic of liquid-phase epitaxial growth of Al_2O_3 crystallites at the S/L interface based on Bolkhovityanov's[21] model.

Molten Al continues to dissolve the bare substrate between the discontinuously precipitated crystallites, resulting in a repetitive, self-sustaining process of ceramic dissolution, increased oxygen solubility in Al(l), and heterogeneous nucleation and epitaxial growth of oxide crystallites on the alumina substrate. Such a growth mechanism is consistent with the model of liquid-phase epitaxial growth due to Y. Bolkhovityanov[21] (Fig. 4).
It has been suggested[16] that epitaxial growth of Al_2O_3 occurs because the first few atomic layers of liquid Al in contact with a monocrystalline substrate are ordered and structurally similar to the substrate (these layers, however, contain excess point defects, which enable O diffusion along the S/L interface toward the droplet center). Thus, it may be conjectured that microcrystalline Al_2O_3 observed at the S/L interface forms by 1) O diffusion along S/L interface, and/or 2) substrate dissolution, O enrichment (or Al–O adsorption), and reprecipitation of alumina.
Another possibility for crystallite growth is the reduced O solubility during droplet cooling and precipitation of Al_2O_3.[17] In general, whatever the dominant process for microcrystalline Al_2O_3 formation, experimental results (Figs. 3(a) to 3(c)) show that more Al_2O_3 crystallites precipitate in the center of the droplet than near the TL; this could possibly be due to a longer S/L contact in the center than near the TL (at TL, ridging occurs at high temperatures[22]). Besides diffusion of O along S/L, convective transport of O to the S/L interface at the droplet center is also possible. At large Peclet numbers, fluid convection will contribute to O transport ($Pe = VR/D$, where V is the fluid velocity, R is the droplet base radius (a few mm), and D is the diffusion coefficient of O in Al; $D \sim 2.2 \times 10^{-10}$ m^2/s at 1373 K [16]). For $R = 2$ mm, $Pe = 9.09V$, where V is in μm/s. For V greater than a few tens of μm/s, fluid convection may play a role.

Table I. Effect of heating procedure on $\theta-\tau_{max}$ relationship

Metal	T	SCH[6,8]		FCH [this study]		
	(K)	θ (°)	τ_{max} (MPa)	θ (°)	τ_{max} (MPa)	τ_{max}^{TC} (MPa)
Al	953	126	19.1	127	63	37.0
	1023	121	29.5	112	40.3	59.2
	1123	96	36.0	110	51.3	31.8
	1223	79	46.3	109	72.0	36.6
	1323	74	48.0	78	54.0	30.6
AlSi11	1023	118	25.4	125	100.5	
	1123	117	60.1	86	102.7	
	1223	85	74.91	77	101.8	
AlTi6	1123	114	17.5	115	15.7	
	1223	90	20.2	92	21.2	
	1323	89	13.5	76	45.3	
AlSn7	953	136	NB			
	1023	123	0*			
	1123	120	0*			
Ti coating	953			116	36[8]	
	1023			113	38[8]	
	1123			45	42[8]	
Sn coating	953			116	22[8]	
	1023			114	26[8]	

τ_{max}^{TC} - shear strength after thermal cycling (200 cycles);
* - failure takes place during cutting; NB - no bonding

Alloying

Table I shows the effect of alloying Al with Ti, Si and Sn at different temperatures on the contact angle, θ, and shear strength, τ_{max}. Alloying Al with 6 wt% Ti slightly increase the θ in Al/Al_2O_3. Higher temperatures decrease the θ, but wetting is achieved only at T>1273 K.

Similarly, AlSn7 alloys show larger θ compared to pure Al. The poor wetting of Al–Sn/Al_2O_3 and Al–Ti/Al_2O_3 is consistent with the literature,[13,23,24] and is due to a non-profitable change in the interface structure and chemistry.

No stable reaction products form in AlTi6/Al_2O_3 under the present test conditions as per the Al–Ti–O phase diagram: this was verified in our SEM/EDS studies that ruled out Ti-rich interfacial phases. In addition, interfacial discontinuities due to shrinkage porosity and cracks (Fig. 5) formed during cooling due to mismatch of coefficients of thermal expansion (CTE) also adversely affect the interface strength. Furthermore, after the dissolution of AlTi6 alloy in HF solution, Al_2O_3 substrate exhibited "smooth" surface profile (Fig. 6), indirectly confirming an absence of any interfacial phases and decreased substrate dissolution.

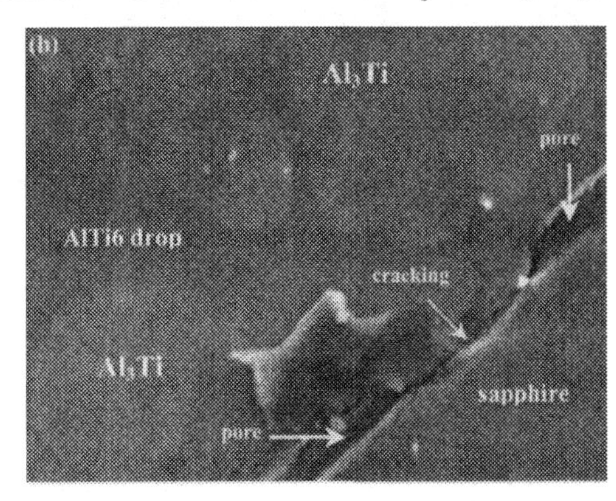

Fig. 5. SEM images of microstructure AlTi6/sapphire interface (1273 K, 4 h) showing discontinuities at the drop/substrate contact due to shrinkage porosity and cracking formed during cooling due to CTE mismatch: (a) periphery of the drop; (b) center of the sample.

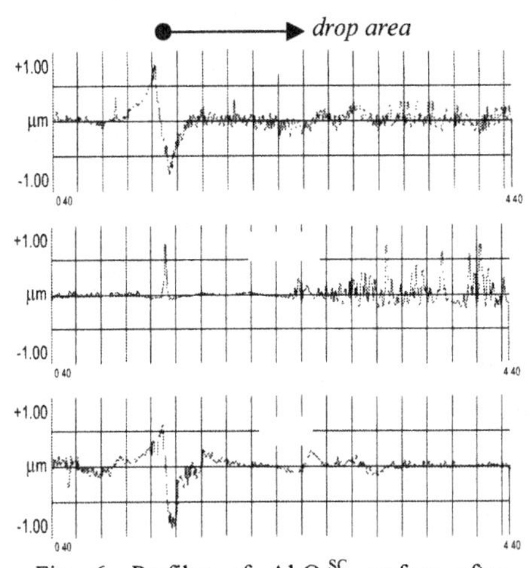

Fig. 6. Profiles of $Al_2O_3^{SC}$ surface after interaction with pure Al, AlSi11 and AlTi6 alloys (1273 K, 4 h) recorded by profilometer after dissolution of the drop in an acid.

An important distinction must be made between how θ and τ_{max} respond to interfacial processes. Whereas the value of θ is determined mainly by the interactions at or near TL, the shear strength, τ_{max}, is determined by the interactions at the S/L interface.

For AlSi11/Al_2O_3 couples (Table I), θ decreases with increasing temperature, approaching 125°, 123°, and 84° at 1073 K, 1123 K, and 1273 K, respectively. At 1273 K ($T>T^w$), there is a strong positive effect of oxide removal on θ. Similarly, the $\theta(CP)$ of AlSi11/Al_2O_3 at 973 K approaches 88° in 18 min; thereafter, θ increases and stabilizes at ~98°. It has been established[13] that α-Al_2O_3 loses O from its basal

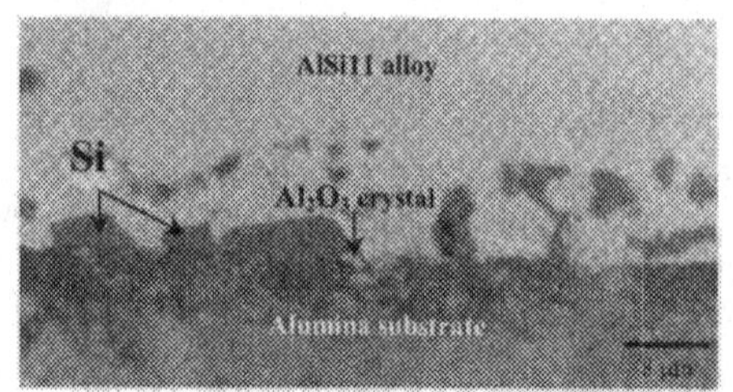

Fig. 7. Microstructure of $AlSi11/Al_2O_3^{PC}$ interface (1273 K, 30 min) showing nucleation of Si crystals at the interface.

(0001) planes during heating in a vacuum to very high temperatures (1500 K), causing surface reconstruction (roughening). However, in the presence of Al and Si vapors, Al_2O_3 surface reconstruction occurs even at 1200 K or lower temperatures[13], thus increasing the solid's surface energy (σ_{sv}), lowering the θ, and enhancing the metal/Al_2O_3 adhesion.

Profilometry of the substrate (Fig. 6) and microstructural examination (Fig. 7) suggest that Si alloying increases the dissolution of Al_2O_3 in the droplet, leading to appreciable roughening. During cooling, Si precipitates at the S/L interface (Fig. 7), where silicon (and the reaction-formed Al_2O_3 crystallites) strengthens the ceramic/metal joint relative to unalloyed Al/Al_2O_3. High Si contents are, however, detrimental to the shear strength as also high Fe contents in the case of Al–Fe/Al_2O_3 (Table II).

Table II. Effect of alloying on θ–τ_{max} relationship (SCH, 1273 K, 30 min)

Metal	θ (°)	τ_{max} (MPa)
Al	77	46.8
AlCu2	133	18.16
AlCu5.65	90	≥71.43*
AlCu32	99	22.81
AlSi5	91	75.55
AlSi11	84	66.19
AlSi20	99	32.44
AlTi0.8	93	30.97
AlTi2	100	33.13
AlTi6	103	17.81
AlFe1	101	53.47
AlFe5	113	29.65
AlFe10	111	0.37
AlFe30	80	6.79

*one sample was broken in ceramic

Fig. 8 shows that mechanical removal of oxide film from AlTi6 drop improves the wetting but not bonding. The strongest effect of mechanical drop surface purification was noted with AlSi11 alloy, i.e., in oxide-free droplets formed near the melting point of Al (973 K), the contact angle and shear strength are comparable to those of the droplet formed at 300 K higher temperature when *in situ* cleaning of the drop from oxide layer takes place.

In the case of Al–Sn/Al_2O_3 (Fig. 9) interfacial shrinkage porosity forms along with a Sn-rich layer on the substrate-side of the AlSn7drop. The Sn-rich layer indicates that sedimentation of the heavier Sn in Al occurs due to density mismatch (ρ_{Sn} ~6710 kg/m^3, ρ_{Al}~2710 kg/m^3

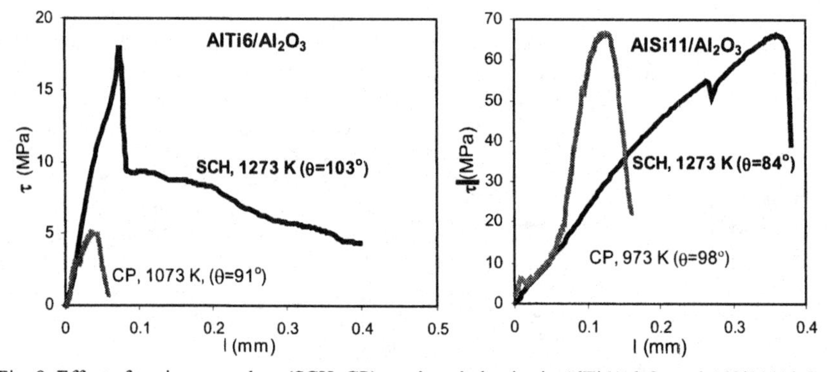

Fig. 8. Effect of testing procedure (SCH, CP) on shear behavior in AlTi6/Al_2O_3 and AlSi11/Al_2O_3.

at 953 K, respectively). As explained in Ref. 8, the Al-rich phase preferentially nucleates inside the drop, but not at the substrate surface, and this leads to shrinkage porosity at the droplet/substrate interface, which weakens the joint, as noted from Table 1. Due to slow contact heating and cooling of Al–Sn/Al_2O_3 couples, and a relatively large freezing range, $\Delta T = (T^{liquidus} - T^{solidus})$ for Al–Sn alloys, the drop remains in a semi-solid state for a long period. Thus, during heating, the eutectic melts first while the residual light Al-rich phase floats to the upper part of the drop, and during cooling, Sn segregates due to large density mismatch. Solidification starts with the formation of α-Al precipitates, which migrate to the upper part of the drop while the heavier (Al–Sn) eutectic settles to the bottom and deposits at the substrate/metal interface.

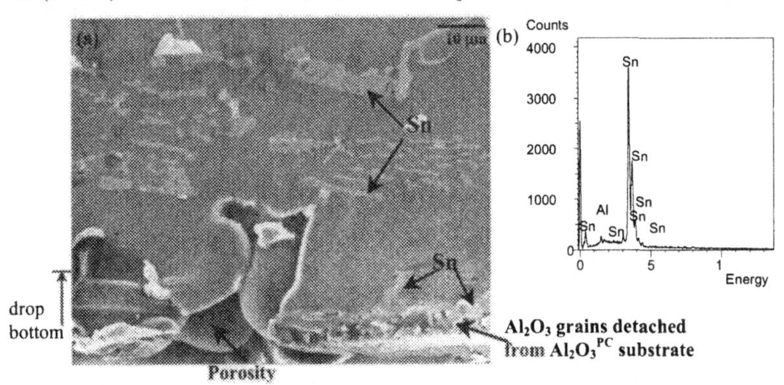

Fig. 9. (a) SEM image of the microstructure of AlSn7 drop (tilt position) debonded from $Al_2O_3^{PC}$ substrate during shear test of AlSn7/$Al_2O_3^{PC}$ couple (1123 K, 30 min); (b) EDS spectrum of the drop bottom showing the presence of Sn-rich phase at the interface.

Substrate Coatings

The effect of Sn and Ti, applied as surface films to Al_2O_3 substrates, on θ and τ_{max} in Al/Al_2O_3 is shown in Table I. Usually, thin (< 1 μm) technological coatings are unstable in contact with molten metals, and it is important to determine if they affect θ and τ_{max}. The data of Table I indicate a clear decrease in θ for Ti-coated Al_2O_3, suggesting that Ti is beneficial as a thin film but not as an alloying element. In fact, Ti film markedly reduces the θ relative to pre-alloyed Ti; in 30 min at 1123 K, θ~45° for Al/Ti/Al_2O_3, and θ~115° for Al–Ti/Al_2O_3.[8] Tin films slightly lower the θ relative to prealloyed Sn; however, for both Al/Sn/Al_2O_3 and AlSn7/Al_2O_3, θ>90°. For example, at 1023 K, with Sn films, an obtuse θ (~114°) is obtained although the Sn film slightly reduces the θ in comparison to pre-alloyed Sn (for which θ~123°).

In Al/Sn/Al_2O_3, Sn film melts (T_m = 504.9 K) during heating and separates into island-like droplets, allowing Al to contact the uncovered regions of the substrate. This is also revealed by the fact that after prolong contact, θ for Al/Sn/Al_2O_3 is similar to that for Al/Al_2O_3. Unlike the Sn film, however, the Ti film in Al/Ti/Al_2O_3 does not melt during heating (T_m = 1933 K); Ti partially dissolves in Al, leading to precipitation of Al_3Ti crystals during cooling on the droplet side of interface. Moreover, during heating to T^{exp} in vacuum even under low Po_2, the interaction of Ti film with Al_2O_3[61] could favorably alter the substrate side of the interface due to formation of a complex Al/Al_2O_3/Ti_xO_y/Ti/Ti_xO_y/Al_2O_3 layer. It appears that the beneficial effect of substrate surface modification using Ti thin films does not fade away.

Interfacial Shear Strength

The values of τ_{max} and θ for various Al alloy/Al_2O_3 couples are summarized in Tables I and II, and Fig. 10. The effect of prealloyed Ti in SCH AlTi6/Al_2O_3 is to reduce the τ_{max} relative to pure Al. The CP AlTi6/Al_2O_3 couples showed a marked drop in τ_{max} at the lower temperatures (1073 K and 1123 K) as compared to the higher temperature (1273 K) SCH couples. Clearly, in the non-reactive AlTi6/Al_2O_3 couples, where prealloyed Ti did not chemically interact with the substrate, oxide removal from the droplet surface in CP procedure did not lead to a stronger joint. Thus, the presence of surface oxide can only diminish the wettability (and joint strength) by obstructing the wettability-enhancing chemical reactions in systems that are inherently reactive, but the oxide removal *per se* does not improve the wettability in non-reactive systems that are inherently non-wettable.

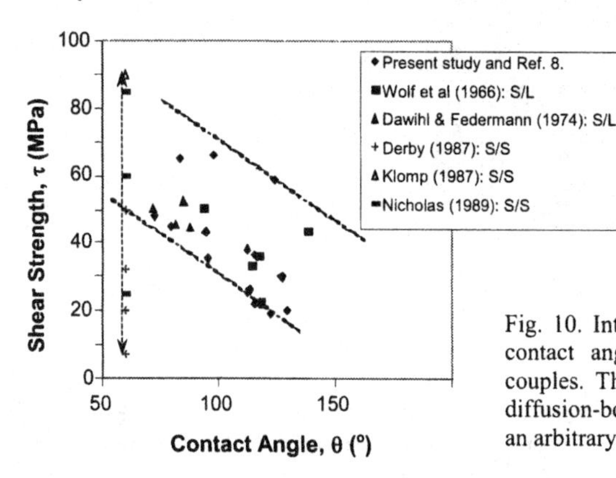

Fig. 10. Interfacial shear strength as a function of contact angle for Al/Al_2O_3 and Al alloy/Al_2O_3 couples. The shear strength values for solid-state, diffusion-bonded (S/S) interfaces are displayed at an arbitrary value of contact angle.

The Ti film increases the τ_{max} in the Al/Ti/Al_2O_3 couples although, as mentioned earlier, pre-alloyed Ti is ineffective. For example, at 1123 K, the τ_{max} is larger for Al/Ti/Al_2O_3 (~42 MPa) than for AlTi6/Al_2O_3 (~17 MPa) and also for pure-Al/Al_2O_3 (~35 MPa). The effect of Sn thin films on τ_{max} is less pronounced than that of Ti. The shear strength data (Table I) on Al/Sn/Al_2O_3 show that tin films marginally improved the τ_{max} at 953 K relative to Al/Al_2O_3 with uncoated substrate; however, at 1023 K, Sn film weakened the interface. In fact, there is a complete lack of wetting and bonding between Sn and Al_2O_3,[23,24] and alloying Al with Sn decreases the W_{ad}, thus weakening the interface. For both Ti and Sn films, τ_{max} increases with the increase in the wettability test temperature, caused by a decrease in contact angle with increasing temperature.

The effect of temperature on θ and τ_{max} is also shown in Table I. For both Al/Al_2O_3 and Al/Ti/Al_2O_3, τ_{max} generally increased with increasing temperature, and the τ_{max} is higher for Al/Ti/Al_2O_3 than for Al/Al_2O_3 at equivalent temperatures. A decrease in θ and an increase in τ_{max} with increasing temperature in Al/Ti/Al_2O_3 are due to formation of wettable reaction products at the joint; however, such products do not form in Al/Al_2O_3 although τ_{max} increases with temperature even at relatively low temperatures at which $\theta > 90°$. The microcrystalline Al_2O_3 that forms at S/L interface (Fig. 3) *via* a dissolution-reprecipitation (or O diffusion) mechanism strengthens the joint even though the measured θ (based on drop profile) might be obtuse. As the quantity of these Al_2O_3 crystallites increases with increasing temperatures (even at $T<T^w$), greater strengthening is realized with increasing temperatures, even when θ remains obtuse.

The shear behavior of the Al–Si/Al_2O_3 couples is quite remarkable in that they exhibit the highest τ_{max} of all the couples (e.g., τ_{max} is 76 MPa and 66 MPa at 5% and 11% Si, respectively). Moreover, τ_{max} is high even at temperatures below T^w (e.g., τ_{max} is ~ 59 MPa at 1073 K for AlSi11), and the effect of CP procedure is less dramatic than in most other couples. This is due to the strengthening contribution of Si crystals that nucleate at the S/L interface (Fig. 11). These Si crystals and the reaction-formed microcrystalline Al_2O_3 strengthen the joint (Tables I and II).

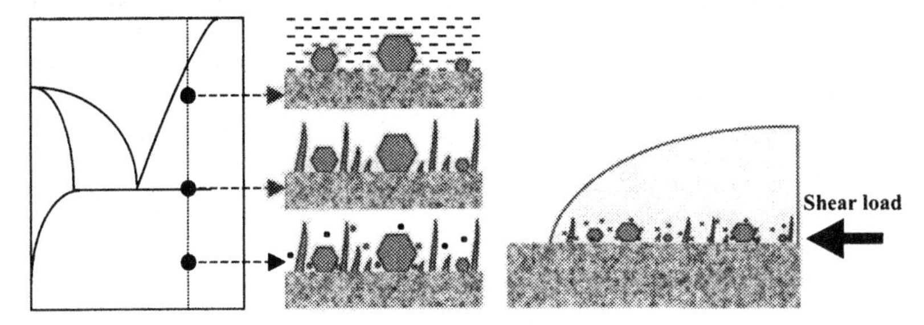

Fig. 11. Schematic of secondary strengthening effect through nucleation of second phase at the interface (in the case of Al–Si, Al–Fe, Al–Cu eutectic systems).

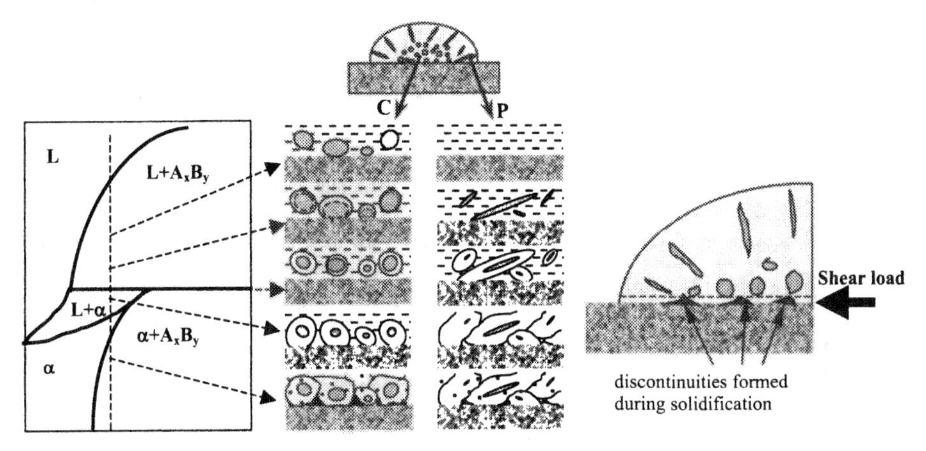

Fig. 12. Schematic of interface weakening due to the formation of shrinkage porosity and cracking in tl center (C) and at the periphery (P) of a drop during cooling caused by CTE mismatch (as in the Al–' peritectic system).

Figure 10 shows a plot of τ_{max} versus θ for some Al alloy/Al_2O_3 couples (literature data are also shown for comparison). In spite of the scatter in the data (due to different composition, substrate characteristics, and test conditions between different studies), it is evident that a small θ results in a large τ_{max} (strong interface). The τ_{max} data[5,19,25-27] in diffusion-bonded couples are shown in Fig. 10 at an arbitrary θ value because measurements of "solid-state" θ are scant. The magnitude of τ_{max} (but not θ) will also be affected by the thermal stresses due to mismatch of thermal expansion coefficient (α) between Al and Al_2O_3. The elastic thermal stress during cooling through an interval, ΔT, is $\Delta\alpha\Delta TE_{Al}$, where E_{Al} is the elastic modulus of Al (~70 GPa), and $\Delta\alpha = (\alpha_{Al} - \alpha_{Al2O3})$. For Al/Al_2O_3, with $\alpha_{Al} = 23.5 \times 10^{-6}/K$ and $\alpha_{Al2O3} = 7.5 \times 10^{-6}/K$, the thermal stresses during cooling from 1123 K and 1073 K to room temperature (298 K) will be ~870 MPa and ~814 MPa, respectively.

These stresses exceed the yield strength of pure Al (~35 MPa), suggesting that extensive interfacial plastic flow of Al on the drop side of the interface may occur. Resistance to deformation will be caused by the microcrystalline Al_2O_3 precipitates, and any hard interface phases (e.g., Si in Al–Si/Al_2O_3 or $CuAl_2$ in Al-Cu/Al_2O_3) that nucleate at the S/L interface as schematically illustrated in Fig. 11 for a eutectic system. On the contrary, if the drop base metal is harder than the second phase precipitates then their presence at the interface may contribute to the interface weakening (e.g., AlSn7/Al_2O_3). However, the effect of second phase precipitates may be negligible if they can be removed from the interface during solidification as it takes place in Al-Ti peritectic alloys (Fig. 12). Moreover, because the interface is strengthened and the thermal stresses are large, some interfacial cracking might also occur (e.g., AlTi6/Al_2O_3 in Fig. 5). Shrinkage porosity formed at the S/L interface (the last region to solidify) will also contribute to the weakening of the interfaces (e.g., with Al-Ti alloys (Fig. 5,12) and with AlSn7 alloy (Fig. 9)).

Effect of Additives in $Al_2O_3^{PC}$

Most oxide-based ceramic additives in $Al_2O_3^{PC}$, such as SiO_2 and TiO_2 react with Al(l). The wetting and shear behaviors of TiO_2 with pure Al are shown in Table III. With the SCH procedure, high wettability test temperatures lower the θ, and θ=64° is obtained at 1373 K in 120 min. The wetting transition temperature, T^w, is between 1173 K (θ=97°) and 1273 K (θ=80°), probably around 1200 K. The SCH Al/TiO_2 samples show a decrease in τ_{max} with increasing wettability test temperature. The effect of CP on τ_{max} is very pronounced, indicating that oxide removal strongly improves the shear strength. It is noteworthy that at 1173 K, θ decreased only slightly (from 96° to 80°) when heating procedure was changed from CH to CP; however, a dramatic increase in τ_{max} occurred at 1173 K when CP was done in place of CH (τ_{max}=42 MPa for SCH and τ_{max}=120 MPa for CP). This suggests that interfacial changes at the S/L interface that determine the joint strength may be revealed in the τ_{max} values than in the θ values, that are sensitive chiefly to the physicochemical processes at TL rather than at S/L interface. TiO_2 is an ionic transition metal oxide with a high exothermic heat of formation (-941 kJ/mol at 1273 K), and a negative Gibbs free energy of formation (-803 to -711 kJ/mol over 773 to 1273 K). Its reaction with liquid Al leads to titanium aluminide formation according to: $TiO_2 + Al \rightarrow Al_2O_3 + Al_xTi_y$, where the type of aluminide formed (e.g., Ti_3Al, TiAl, Al_3Ti) depends upon the ratio of TiO_2 to Al. Thus, improvement in wetting in Al/TiO_2 can be attributed to favorable interface structure resulting from chemical reactions, when the negative effect of surface oxides is eliminated by CP.

Table III. θ–τ_{max} relationship in Al/TiO_2

Conditions	θ (°)	τ_{max} (MPa)
SCH, 1173 K, 2h	96	45
SCH, 1273 K, 2h	80	35
SCH, 1373 K, 2h	64	25
CP, 1173 K, 2h	80	121

Wetting tests on common additives to alumina such as SiO_2, CoO, NiO, ZnO, TiO_2, ZrO_2, mullite, and fly ash show two fundamental types of interactions[28,29]: reactive metal *penetration* of dense polycrystalline ceramics, and reactive *infiltration* of porous ceramics. In both reactive penetration and reactive infiltration, chemical reactions form a C^4 structure (i.e., co-continuous ceramic composite), which is essentially an interpenetrating network of Al and Al_2O_3. The formation of a C^4 structure in porous ceramics by infiltration arises from reaction-enhanced capillary-driven flow through pores, while in the case of polycrystalline ceramics – by reactive penetration of molten Al through grain boundaries. However, C^4 structure also forms in the substrates free of grain boundaries such as single crystals (e.g., ZnO^{SC}, NiO^{SC}, CoO^{SC}) and amorphous SiO_2 or mostly amorphous fly ash. Interestingly, Al penetrates and forms a C^4 structure in non-wetting dense ceramics such as ZnO^{SC}, which are not wet by Al at $T \leq 1273$ K, but Al does not penetrate wettable ceramics such as TiO_2 and ZrO_2 even when they are porous. In spite of high wettability of NiO^{SC} and CoO^{SC} by molten Al reactive metal flow does not occur in these materials; instead, just a 'dispersion' of the reaction product layer in the drop is noted.

The formation of a C^4 structure in dense ceramics and a lack of reactive metal flow through wettable porous ceramics have been attributed to the cracking of the substrate and freshly formed alumina. For example, substrate cracking due to thermal expansion mismatch and specific volume changes accompanying the reaction could cause flow cessation in dense SiO_2 substrates. Other possibilities are the secondary oxidation of metal front, and increased resistance to penetration from epitaxial alumina crystallites. Secondary oxidation could prevent the reactive spreading on SiO_2 that will impair the wetting. The oxygen needed for secondary oxidation could be from localized substrate dissolution, or from diffusion along the S/L interface. SiO_2 will then locally dissolve in the liquid, resulting in volume expansion and microcracking, but secondary oxidation of the front will prevent metal penetration.

Another possibility for the C^4 structure formation is the occurrence of a eutectic reaction in the Al–O system. The C^4 structure of the Reaction Product Region (RPR) formed between Al and dense reactive oxides such as single crystals ZnO^{SC}, NiO^{SC} and CoO^{SC} as well as amorphous SiO_2 or mostly amorphous fly ash substrates consists of a characteristic eutectic-like (interpenetrating) appearance. In fact, the possibility of an actual eutectic structure cannot be ruled out especially in view of the suggestions[30] made in the literature concerning the existence of eutectic transformation on the Al–O phase diagram. There are well-known examples of eutectic reactions in Mo–C (e.g., Mo+Mo_2C eutectic), Nb–C, Ta–C, V–C, W–C, and U–C. However, direct experimental confirmation of the Al–O eutectic has been lacking.

The RPR plays a pivotal role in the propagation of the spreading front. Sessile-drop experiments on Al/SiO_2 couples[28], with four thin SiO_2 layers (0.3 mm thick) stacked on top of one another, were done to examine the role of RPR in spreading front advance. It was noted that cracking of SiO_2 near the triple line (due to expansion accompanying $SiO_2 \rightarrow Al_2O_3$ transformation) did not hamper RPR propagation. The RPR covered a wider surface area on the top SiO_2 strip compared to the situation when thicker SiO_2 substrates were used. Interestingly, the RPR did not penetrate beyond the 0.3 mm thick first silica layer. Thus, the sideways (lateral) growth of RPR was not hampered by surface cracking resulting from reduction of SiO_2 by Al, and the RPR advanced beyond the triple line.

The influence of additives and impurity in polycrystalline Al_2O_3 on metal/Al_2O_3 bonding could be substantial. For example, $Al_2O_3^{PC}$ substrates having even small amount of SiO_2 (less than 1%) bond stronger to Al droplets than SiO_2-free substrates, and our TEM examination shows that the SiO_2 vigorously reacts with Al to form needle-like Si precipitates at the interface, which reinforce (strengthen) the interface (frequently, however, methodological difficulties in phase identification have led researchers to erroneously label the Si+Al_2O_3 complexes, composed of Si needles and epitaxial alumina crystallites, as mullite). On the other hand, additives such as SiO_2, mullite and kaolin are less reactive to Al than NiO, CoO and iron oxides, but more reactive than TiO_2 and ZrO_2. Investigations on the influence of oxide ceramics on the reactive metal penetration along the grain boundaries in polycrystalline composite ceramics such as Al_2O_3–SiO_2, Al_2O_3–NiO, and Al_2O_3–NiO are needed to understand how the response of individual component oxides affects the reactivity, wettability, and shear strength.

CONCLUSIONS

The wetting-interface strength relationship in high-temperature ceramic/metal couples must be interpreted in light of nano- and micro-scale structure of the interface. New experimental results on the effect of liquid-phase ceramic joining parameters (time, temperature, alloying, ceramic additives, and metal films on ceramic) on the wetting-structure-strength response of Al/Al_2O_3 and Al/TiO_2 couples suggest that non-equilibrium phenomena (segregation, sedimentation, dissolution, and defects) markedly influence the interface behavior. It is argued that data on the classical liquid-state joining science parameters (contact angle and work of adhesion) must be coupled with the structural information to develop a scientific understanding of the joining process.

ACKNOWLEDGEMENTS

This work was performed at the Foundry Research Institute (Poland) in cooperation with the University of Wisconsin-Stout under the project No. 7T08B 00320 and No. IOd 8029/00 supported by the State Committee for Scientific Research in Poland. Additional support was provided by the U.S. National Academy of Sciences under the COBASE program (Contract No. INT-0002341) from the U.S. National Science Foundation. The authors thank W. Radziwill, M. Ksiazek and B. Mikulowski for technical assistance.

REFERENCES

[1] N. Sobczak, M. Ksiazek, W. Radziwill, J. Morgiel and L. Stobierski, "Effect of Titanium on Wettability and Interfaces in the Al/SiC System"; pp. 138–144 in *Reviewed Proc. Second Int. Conf. „High Temperature Capillarity"*, edited by N. Eustathopoulos and N. Sobczak, Foundry Research Institute, Poland, 1998.

[2] W.H. Sutton, *Report R-64 SD44, GE Space Sciences Lab*, U.S.A. (1964).

[3] M. Nicholas, "The strength of metal/alumina interfaces", *J. Mater. Sci.*, **3**, 571–576 (1968).

[4] M. Nicholas, R.R.D. Forgan, D. M. Poole, "The adhesion of metal/alumina interfaces", *J. Mater. Sci.*, **3**, 9–14 (1968).

[5] S. M. Wolf, A.P. Levitt, and J. Brown, "Whisker-metal matrix bonding", *Chem. Eng. Progress*, **62** [3] 74-78 (1966).

[6] N. Sobczak, R. Asthana, M. Ksiazek, W. Radziwill, B. Mikulowski and I. Surowiak, "Influence of Wettability on the Interfacial Shear Strength in the Al/alumina System"; pp. 129–142 in *State of Art in Cast Metal Matrix Composites in the Next Millennium*, edited by P.K. Rohatgi, TMS Publications, Pennsylvania, USA, 2000.

[7] N. Sobczak, "Wettability, Structure and Properties of Al/Al_2O_3 Interfaces", *Kompozyty*, **3** [7], 301–312 (2003).

[8] N. Sobczak, R. Asthana, M. Ksiazek, W. Radziwill, and B. Mikulowski, "The effect of temperature, matrix alloying and substrate coatings on wettability and shear strength of Al/Alumina couples", *Metall. Mater. Trans.*, **35A** [3], 911–924 (2004).

[9] N. Sobczak, L. Stobierski, M. Ksiazek, W. Radziwill, J. Morgiel and B. Mikulowski, "Factors Affecting Wettability, Structure and Chemistry of Reaction Products in Al/Si_3N_4 System", *Transactions of Joining and Welding Research Institute*, **30**, 39–48 (2001).

[10] N. Sobczak, M. Ksiazek, W. Radziwill, L. Stobierski and B. Mikulowski, "Wetting-Bond Strength Relationship in Al–AlN System", *Transactions of Joining and Welding Research Institute*, **30**, 125–130 (2001).

[11] N. Sobczak, K. Nogi, H. Fujii, T. Matsumoto, K. Tamada, and R. Asthana, "The effect of Cr thin film on wettability and bonding in Ni/Alumina couples", pp. 108–115 in *Joining of*

Advanced & Specialty Materials, Edited by J. N. Indacochea *et al.*, ASM International, Materials Park, OH, 2003.

[12]M. Ksiazek, N. Sobczak, W. Radziwill, and B. Mikulowski, "Influence of surface modification of alumina substrates on wetting-bond strength relationship in Cu/alumina", pp. 96–100 in *Joining of Advanced & Specialty Materials*, Edited by J. N. Indacochea *et al.*, ASM International, Materials Park, OH, 2003.

[13]Eustathopoulos, M.G. Nicholas and B. Drevet, *Wettability at High Temperatures*, Pergamon, 1999.

[14]D.A. Weirauch and W.J. Krafick, "The effect of carbon on wetting of aluminum oxide by Al", *Metall. Trans.*, **21A**, 1745–1751 (1990).

[15]W.D. Kaplan, "Al-alumina interfaces: Non-equilibrium wetting in a binary system", pp. 153-160 in *Interfacial Science for Ceramic Joining*, Edited by A. Bellosi, T. Kosmak and A.P. Tomsia, vol. 58, NATO ASI Series, High Technology, Kluwer Academic Publishers, 1998

[16]G. Levi and W.D. Kaplan, "Oxygen-induced interfacial phenomena duing wetting of alumina by liquid aluminum", *Acta Mater.*, **50**, 75–88 (2002).

[17]P.D. Ownby, K.W.K. Li, and D.A. Weirauch, Jr., "High-temperature wetting of sapphire by aluminum", *J. Amer. Ceram. Soc.*, **74** [6] 1275–81 (1991).

[18]V. Laurent, D. Chatain, C. Chatillon, and N. Eustathopoulos, "Wettability of monocrystalline alumina by Al between its melting point and 1273 K", *Acta Mater.*, **36** [7] 1797–1803 (1988).

[19]M. Nicholas, "Interactions at oxide-metal interfaces", *Materials Science Forum*, **29**, 127–150 (1988), Trans Tech, Switzerland.

[20]V.M. Perevertailo, S.M. Samokhin, O.G. Kulik, and M.A. Opanasenko, "Peculiarities of the contact interaction of dispersion-hardened copper with adhesion-active metals and kinetic characteristics of their spreading and wetting", *J.l of Superhard Materials*, **24** [1] 28–32 (2002).

[21]Y. Bolkhovityanov, "Growth instabilities in LPE processes: the initial steps of heteroepitaxy", pp. 37–53 in Epitaxial Crystal Growth, Edited by E. Lendvay, Trans Tech Publications, Switzerland, 1991.

[22]E. Saiz, R.A. Cannon and A.P. Tomsia, "Energetics and atomic transport at liquid metal/alumina interfaces", *Acta Mater.*, 47 [15] 4209–4220 (1999).

[23] I. Rivollet, D. Chatain and N. Eustathopoulos, "Wetting behavior of single crystal alumina with gold and tin between their point of fusion and 1673 K", *Acta Metall.*, **35**, 835–844 (1987).

[24]J.G. Li, L. Coudurier, and N. Eustathopoulos, "Work of adhesion and contact angle isotherm of binary alloys on ionocovalent oxides", *J. Mater. Sci.*, **24**, 1109-1116 (1989).

[25]W. Dawihl and H. Federmann, "Einflub metallischer aufdampfschichen auf das benetzungsverhalten von reinstaluminium auf Al_2O_3–unterlagen", *Aluminium*, **50** [9] 574–577 (1974).

[26]B. Derby, "The influence of surface roughness on interface formation in metal/ceramic diffusion bonds", *Materials Science Research*, **21**, 319–328 (1987).

[27]J.T. Klomp, "Solid-state bonding of metals to ceramics", *Sci. Ceram*, **5**, 501–522 (1970).

[28]N. Sobczak, "Wettability and Reactivity between Molten Aluminum and Selected Oxides", to be published in *Proc. E-MRS Fall Meeting, Symposium G: Bulk and Graded Materials*, Warsaw, Poland (2003).

[29]E. Saiz and A.P. Tomsia, "Kinetics of metal-ceramic composite formation by reactive penetration of silicates with molten Al", *J. Amer. Ceram. Soc.*, **81** [9] 2381–2393 (1998).

[30]ASM Engineered Materials Reference Book, ASM International, Ohio, USA, 1989, p. 199.

INTERFACE STRUCTURES AND DIFFUSION PATHS IN SiC/METAL COUPLES

Masaaki Naka, Takashi Fukai
Joining and Welding Research Institute, Osaka University
11-1 Mihogaoka, Ibaraki, Osaka 567, Japan

Julius C. Schuster
Department of Physical Chemistry, University of Vienna
Waheringer Strasse 42, A1050 Vienna, Austria

ABSTRACT

The interface structures between SiC and metal are reviewed in SiC/metal systems. Metals are divided into two groups, carbide forming metals and non-carbide forming metals. Carbide forming metals form metal carbide on the metal side, and metal silicide on the SiC side. Further diffusion of Si and C from SiC causes the formation of a ternary phase. Non-carbide forming metals form a metal silicide containing graphite or a layered structure consisting of metal silicide and metal silicide containing graphite. The diffusion path between SiC and metal are formed along tie-lines connecting SiC and metal on the corresponding ternary Si-C-M phase diagram. The reactivity of metals is dominated by the formation of carbide or silicide. The reactivities of elements are discussed relative to their positions in the periodical table of elements, and Ti shows the highest reactivity among carbide forming metals. For non-carbide forming metals the reactivity sequence is Fe>Ni>Co.

INTRODUCTION

Silicon carbide is a candidate for high temperature structural components because of its remarkably light weight and high strength. SiC to metal bonding techniques are necessary to expand the engineering application of the ceramics. Detailed knowledge of phase reactions and diffusion paths for SiC/metal systems are needed in order to control the interface structures in SiC/metal joints at high temperatures. In this work, the formation of reaction products and interface structures in SiC/metal systems are reviewed and discussed along with the diffusion paths of corresponding Si-C-Metal systems.

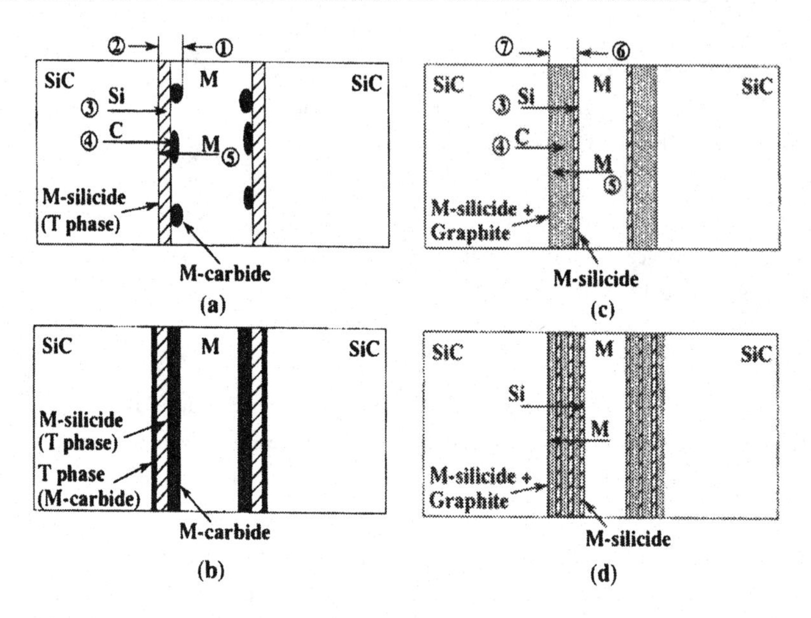

Fig. 1 Schematic structure of the interface reaction and diffusion of elements in SiC/metal couples. (a) initial stage for carbide forming element, (b) intermediate stage for carbide forming elemnt, (c) initial stage for Fe, and (d) initial stage for Co and Ni.

Initial Stage of Reaction Process

In initial reactions at the interface between SiC and metals, SiC decomposes to Si and C, and forms metal silicides, metal carbides and ternary carbosilicides according to the following equations, dependant on the elements involved.

$$M + C \rightarrow \text{M-carbide} \quad (1)$$

$$M + Si \rightarrow \text{M-silicide} \quad (2)$$

$$M + Si + C \rightarrow \text{T-ternary phase} \quad (3)$$

$$M + Si + C \rightarrow \text{M-silicide} + \text{Graphite} \quad (4)$$

In the case of elements in IVa (Ti, V, Hf), Va(V, Nb,W) and IVa (Cr, Mo, W) granular carbides appear, and grow to form a carbide zone on the metal side. A silicide zone or ternary phase zone appears and grows on the SiC side (Figs. 1a and b). Metals (Fe, Co, Ni) in VIII react with SiC to form metal silicides and graphite. In Fe/SiC system graphite distributes in the silicide, and M-silicide and graphite form layered structure in Co and Ni/SiC systems,. because C is immobile and Si and metal mutually diffuse through the graphite zone (Figs. 1c).

Intermediate Stage of Reaction Process

In the case of elements in IVa (Ti, Zr, Hf), Va (V, Nb, Ta) and Via (Cr, Mo, W) the following reactions processes take place. ① Carbon diffuses through the reaction zone and granular carbides form a carbide zone on the metal side (Fig.1b). ② In the SiC /reaction zone interface Si and C diffuse from the SiC side to the metal side and metal diffuses to the SiC side. Furthermore, M-silicide grows or a ternary phase appears due to the diffusion of large amounts of Si and C from SiC (Fig. 1b) .

The growth rate of metal silicide and carbide decelerate due to the formation of the ternary phase.

Table 1 Diffusion path in the SiC-Metal systems.

	Systems	Diffusion path in SiC/ Metal systems		ref.
IVa	SiC/ Ti	SiC/ Ti_3SiC_2 / $Ti_5Si_3C_x$+TiC / TiC / Ti	(1373K-1773K)	1
	SiC/ Zr	SiC / ZrC / $Zr_5Si_3C_x$ / Zr_2Si / ZrC_x / Zr	(1473K- 1673K)	2
	SiC/ Hf*	SiC / HfC / $Hf_5Si_3C_x$ / Hf_2Si / HfC_x / Hf		3
Va	SiC/ V	SiC / $V_5Si_3C_x$ / V_3Si / V_2C + V / V	(1473K-1673K)	4
	SiC/ Nb	SiC/ NbC / $Nb_5Si_3C_x$ / Nb_5Si_3/ Nb_3SiC_x/ Nb_2C/ Nb	(1473K)	5
		SiC/ NbC / $Nb_5Si_3C_x$/ NbC/ Nb_2C/ Nb	(1790K)	
	SiC/ Ta	SiC/ TaC / $Ta_5Si_3C_x$ / Ta_2C / Ta	(1773K)	6
VIa	SiC/ Cr	SiC/ $Cr_5Si_3C_x$/ Co_3SiC_x/ Cr_7C_3/ $Cr_{23}C_6$/ Cr	(1373K-1673K)	7
	SiC/ Mo	SiC/ $Mo_5Si_3C_x$/ Mo_5Si_3 / Mo_2C / Mo	(1473K)	8
	SiC/ W*	SiC/ W_5Si_3 / W_2C / W		3
VIII	SiC/ Fe	SiC/ Fe_3Si+C_G/ Fe_3Si / Fe	(1123K- 1373K)	9
	SiC/Co	SiC/ Co_2Si+C_G / Co_2Si / Co(Si) / Co	(1273K)	9
		SiC/ CoSi+C_G / Co_2Si+C_G /Co_2Si / Co(Si) / Co	(1423K)	
	SiC/ Ni	SiC/ Ni_2Si+C_G/ $Ni_{31}Si_{12}$+C_G/ Ni_3Si+C_G/ Ni_3Si/ Ni	(1123K-1323K)	9

* Presumpsion from equilibrium diagram

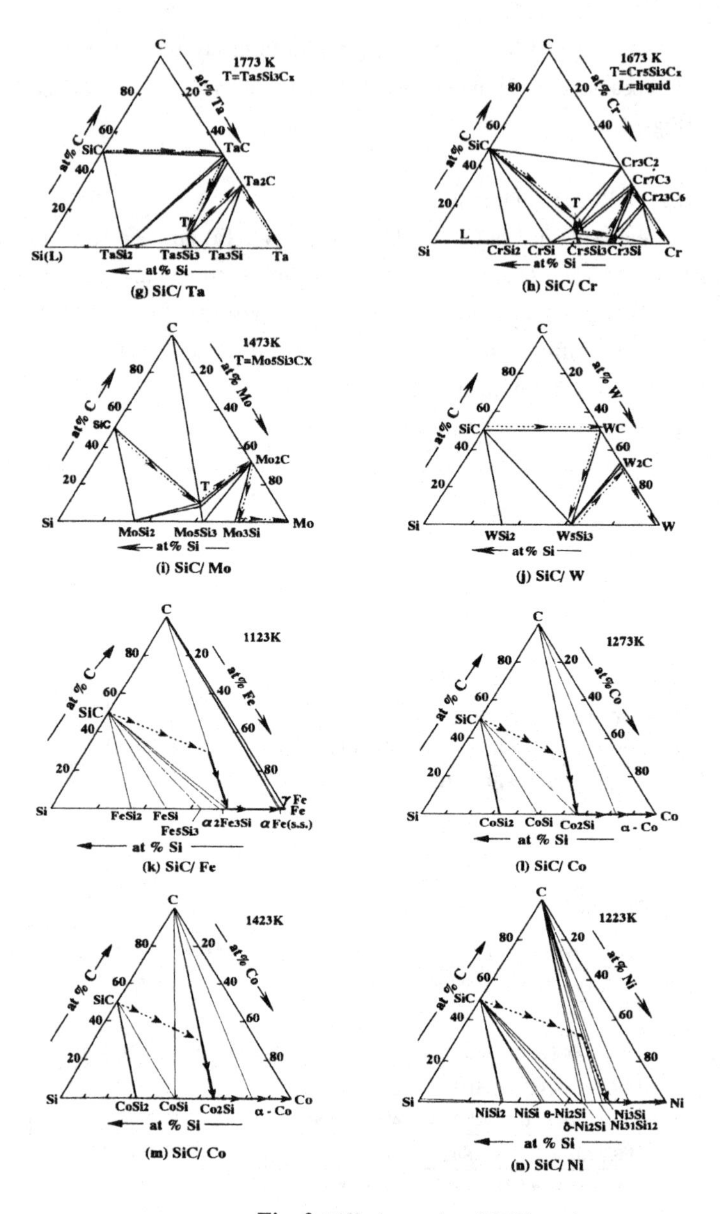

Fig. 2 Diffusion paths of SiC/metal systems

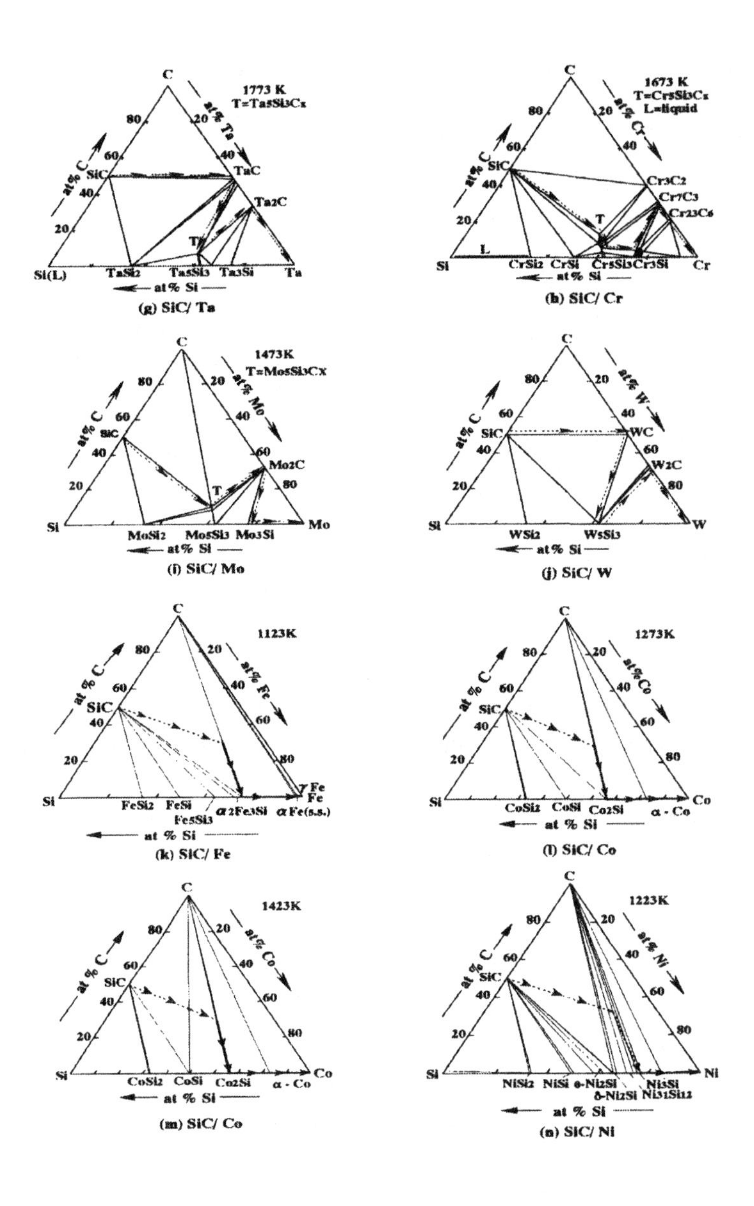

Fig.3 Diffusion paths of SiC/metal systems.

In the case of elements in VIII (Fe, Co, Ni), a new phase does not appear. The silicide zone containing graphite grows with reaction time in the Fe/SiC system, while a layered structure of silicide and graphite grows in Co, Ni/systems (Fig. 1d).

The following general rules dominate the reactions between SiC and metal.

(1) At constant temperature the diffusion path between SiC and metal appears after a reaction time and is described on the corresponding Si-C-M system. The diffusion path then disappears after the metal has been consumed.

(2) The diffusion path depends on the reaction temperature and changes when the corresponding phase diagram changes (as in the case of SiC/Nb and SiC/Co).

(3) The diffusion path follows a straight line connecting SiC and metal according to the material balance.

(4) As the metal thickness changes, the diffusion path doesn't change though the apparent time of reaction and phase changes depend on the thickness.

The observed diffusion paths of SiC and metals are listed in Table 1, and are expressed on the corresponding phase diagram of Si-C-M shown in Figs. 2 and 3 .

Kinetics of growth for reaction phases

In a SiC/metal couple the consumption of metal takes place during the final reaction stage and Fick's law doesn't dominate the reaction process. The reaction process of a SiC/metal couple is divided into a first stage at which some metal remains and a second stage in which the metal is consumed and equilibrium is attained. In the SiC/metal couple the growth of reaction zone is dominated by Fick's law,

$$x^2 = k \cdot t \qquad (1)$$

$$k = k_0 \exp(-Q/RT) \qquad (2)$$

where x, k, k_0, and Q are thickness of reaction zone, rate constant, and activation energy for growth of reaction zone, respectively. The total thickness of reaction zones in SiC/metal couples is dominated by eqs. 1 and 2. during the first stage. The rate constant of SiC/metal are summarized in Fig. 4, where the metal is Ti, V, Cr, Fe, Co, Ni, Zr, Nb, Mo, of Ta. The k of the SiC/metal is plotted as a function of 1/T. At the reaction condition of 1573K and 14.4ks, the reaction thicknesses of SiC/metal (M= Ti, Zr, V, Nb, Ta, Cr, Mo) are Ti=30, Zr=5.6, V=16, Nb=0.6 and Ta=0.5μm. At the reaction condition of 1273K and

3.6ks the reaction thicknesses of SiC/metal (M=Fe, Co, Ni) are Fe=142, Co=18 and Ni=44 μm. The reactivities of Ti, V, and Cr which easily form metal carbides and silicides are relatively higher than those of other metals. In metals belonging to a single same group (IVa), the reactivates are V>Nb>Ta. The reactivities of metals in group VIII are Fe>Ni>Co.

The reactivity of metal in contact with SiC is discussed with the periodical table shown in Fig.5. The reactivity of metal can be directly related to the thickness of the reaction zone formed at a constant temperature and time. In the same group, the reactivity of metals with lower atomic weights are higher. These trends of reactivities for metals may be related to the ease of formation of metal carbides and silicides and ternary phases. In Fig 5 the reactivities of carbide-forming metals are expressed by the number attached, and

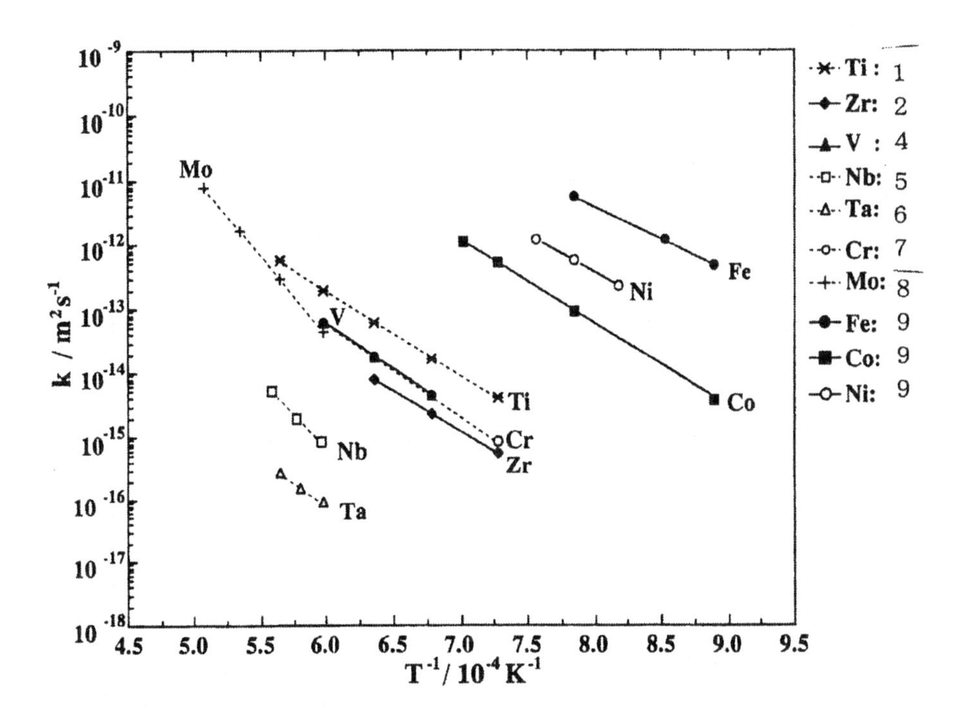

Fig. 4 Rate constants of SiC/metal systems.

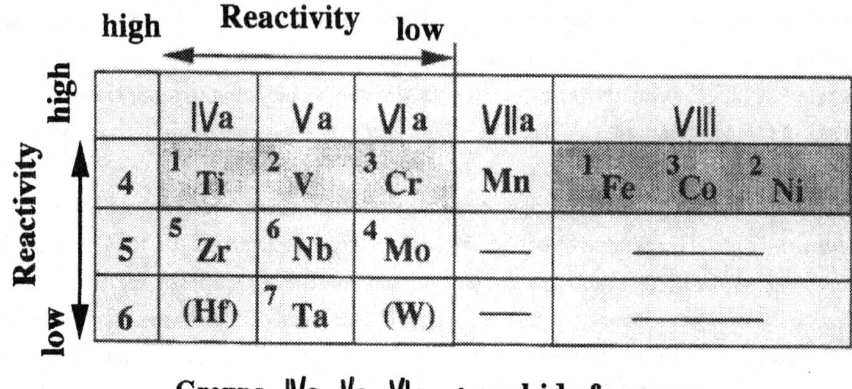

Fig. 5 Reactivates of elements on the periodical table.

Ti shows the highest reactivity. Among non-carbide forming metals Fe shows the highest reactivity.

SUMMARY

In SiC/metal couples at high temperatures the reaction structures and phases formed at the SiC/metal interface are reviewed. Metals are divided into carbide forming elements and non-carbide forming elements based on their reactivity with Sic.

With carbide forming metals, metal silicide zones and metal carbide zones are formed at the SiC/metal interface during an initial reaction period. Depending on the metal, a ternary phase zone is formed by further diffusion of Si and C at the interface between SiC and metal silicide.

Among non-carbide forming elements, silicide zones containing graphite form for Fe. Meanwhile, silicide and silicide containing graphite layered zones for Co or Ni are formed during the initial reaction time.

The reactivity of metal in contact with SiC is discussed in terms of the kinetics of reaction zones growth. Ti yields the highest reactivity, and the reaction zone in SiC/Ti quickly grows at high temperatures. For carbide forming elements which belong to the same group, metals with lower atomic weights yield the higher reactivity. For non-carbide forming elements in group VIII, the sequence of reactivity is Fe, Co and Ni. The reactivity of elements in contact with SiC depends on the carbide and silicide forming abilities.

REFERENCES

[1] M. Naka, J. C. Feng and J. C. Schuster, Metall. And Mater, Trans. A, 28A(1977), 1385-1390.

[2] T. Fukai, M. Naka and J. C. Schuster, Trans. Join. Weld. Res. Inst., 25(1996), 59-62.

[3] J. C. Schuster, Structural Ceramic Joining II, Ceramic Trans. 35(1993), 43-57.

[4] T. Fukai, M. Naka, J.C.Scuster, J. Japan. Inst. Metals, 62-10(1998), 899-904.

[5] J. C. Feng and M. Naka, J. Japan Inst. Metals, 59-1(1997), 456-461.

[6] J. C. Feng, M. Naka and J. C. Feng, J. Japan Inst. Metals, 61-5(1997), 636-642.

[7[J. C. Feng, M. Naka and J. C. Feng, J. Japan Inst. Metals, 61-7(1997), 636-642.

[8] F. J. J. Van Loo, F. M. Smet, G. D. Riek, High Temp. SSi. High Pressures, 14(1982), 25-31.

[9] T. Fukai, PhD Thesis, Osaka Univ., 1999, Dec.

PHOTOCATALYTIC TITANIA COATINGS BY A LOW-TEMPARATURE SOL-GEL PROCESS

Walid A. Daoud* and John H. Xin
Nanotechnology Centre,
Institute of Textiles & Clothing,
The Hong Kong Polytechnic University,
Hung Hom, Hong Kong

ABSTRACT

Transparent thin film coatings of sol-gel derived titanium dioxide were produced at low temperatures. The photocatalytic activity of the formed layers was studied by means of their antibacterial property. The UV-Vis transmission of the coatings was investigated. The effect of film coating on the mechanical and physical properties of the substrates is discussed.

INTRODUCTION

Crystalline titanium dioxide has received increasing attention due to its interesting properties and potential applications, e.g. photocatalysts,[1-3] photovoltaics,[4] gas sensors,[5] and electrochromic display devices.[6-7] In recent years, great interest in the photocatalytic activity of the anatase form has been growing and the bactericidal activity of anatase TiO_2 photocatalyst in aqueous media has been reported.[8] Several papers have discussed the electrical properties of anatase titania thin films prepared using relatively high temperature deposition methodologies, such as chemical vapor deposition (CVD)[9] and sputtering.[10]

Recently, the sol-gel process and dip-coating technique were employed to produce porous metal oxide thin films.[11-13] However, high temperatures of 500-550°C were used to produce those films. The formation of photocatalytic titania films at low temperature is important for the fabrication of transparent films on substrates that can not withstand high temperature treatment, such as plastics and textiles. High temperature treatment would limit the utilization of the optical and photocatalytic properties of sol-gel derived titania in these application areas. In addition, porous coating may not a have high level of washfastness due to the peel-off effect. Washfastness is a requirement for textiles in particular.

It is interesting to note that there is a relation between the absorption intensity of UV radiation and the activity of the catalysts.[14] The stronger the UV-absorption intensity, the higher the activity, where the strong absorption intensity implies that more electrons can be promoted from the valence band into the conduction band and more separate electrons or holes can be produced, which will help to enhance the photocatalytic activity. The sol-gel technique employs either inorganic sol precursors[15] or organometallics such as metal alkoxides which are usually soluble in alcohol.[16] In this study, the second method was employed to form strongly UV-absorbing photocatalytic titania thin coatings on cotton fabrics using a low temperature process.

* Corresponding author, tcdaoud@polyu.edu.hk

EXPERIMENTAL

The nanosol was prepared at room temperature by mixing titanium tetraisopropoxide (Aldrich, 97%) with absolute ethanol (Riedel, 99.8%) at pH of 1 to 2. The mixture was vigorously stirred for 10 minutes prior to coating. A 10 x 10 cm knitted cotton substrate was dried at 100°C for 30 minutes, dipped in the nanosol for 30 seconds and then padded using an automatic padder at a nip pressure of 2.75 kg/cm^2. The padded substrates were then dried at 80°C for 10 minutes in a preheated oven to drive off ethanol and finally cured at 150°C for 5 minutes in a preheated curing oven.

The structure of the titania coatings was studied by scanning electron microscopy using a Leica Stereoscan 440 equipped with an Oxford Energy Dispersive X-ray System, operating at 20 kV. The washing was carried out following AATCC test method 61-1996[17] using AATCC Standard Instrument Atlas Launder-Ometer LEF. The UV-Vis transmission of the coated fabric was performed according to the Australian/New Zealand Standard AS/NZS 4399:1996 using Varian Cary 300 UV/Vis Spectrophotometer. The antibacterial assessment was conducted following AATCC test method 147-1998.[17] X-ray diffraction analysis was performed using a Bruker D8 Discover X-ray Diffractometer operating at 40 kV. The bursting strength test was carried out according to the ASTM D3786-01 Standard Test Method for Hydraulic Bursting Strength of Textile Fabrics.[18] The tensile strength test was carried out according to the ASTM D5025-95 Standard Test Method for Breaking Force and Elongation of Textile Fabrics.[18] An air permeability study of the coated fabrics was conducted on an air permeability tester (Kato Tech Co Ltd, Kyoto, Japan). In this study, a constant air flow is generated and passed through the fabric specimens and the air resistance is measured by the loss of air pressure.

RESULTS AND DISCUSSION

In contrast to the SEM image of a cotton fiber (Fig. 1a), the surface of the titania-coated cotton fiber was smoother which indicates the formation of a uniform continuous layer (Fig. 1b). Fig. 1c shows that the shape of the particles is near spherical and the average particle size is about 20 nm in diameter.

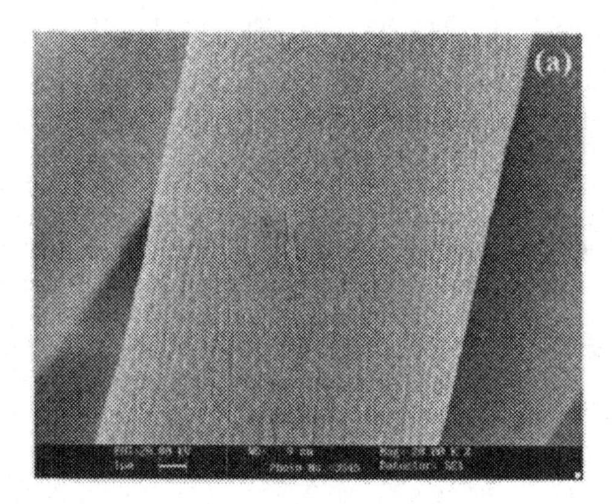

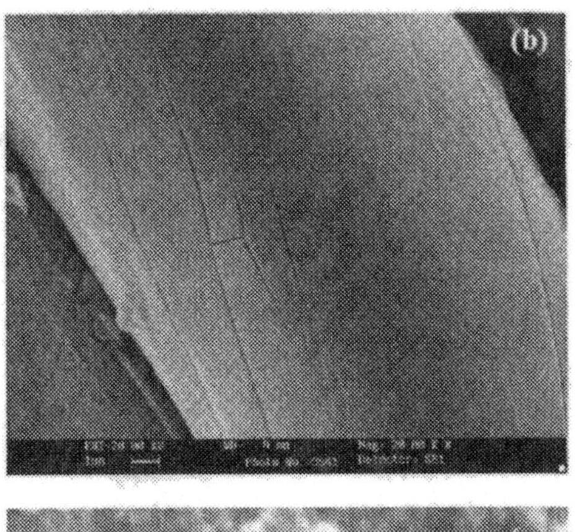

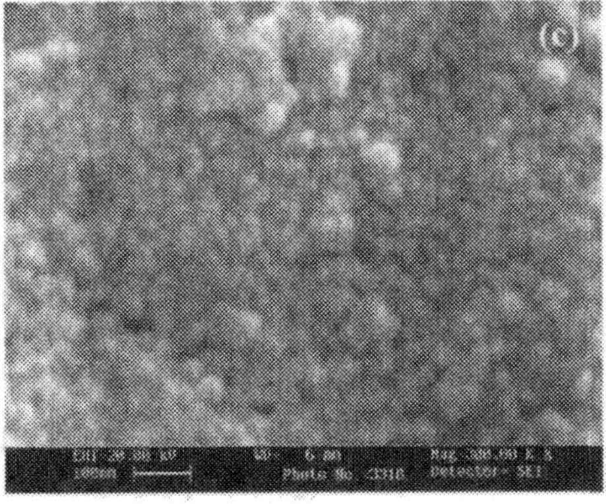

Figure 1. SEM images of (a) cotton fiber, (b) titania coated cotton fiber, showing the morphological change in the surface structure, (c) higher magnification image of titania coated cotton, revealing the nanostructure and particle size of the titania coating.

The UV absorption study of the coated fabrics (Fig. 2a) revealed a high 50+ UPF (UV Protection Factor) rating, an excellent protection classification according to the Australian/New Zealand Standard, compared to a non-ratable UPF rating (10) for the uncoated fabric (Fig. 2b). In the visible light region, the transmission of the coated fabric was about 10-15% lower than that of the uncoated fabric. At 496 nm wavelength, a linear drop of transmission occurred until 332 nm in the UV region, at which wavelength a complete blocking of UV wavelengths (from 332 to 280 nm) was observed. The UV-blocking and 50+ UPF characteristics were reproducible after 20 home launderings (Fig. 2c) indicating good adhesion between titania and cotton. This is considered to be due to covalent bonding resulting from a dehydration reaction between the hydroxyl groups of cotton and the hydroxyl groups of titania.

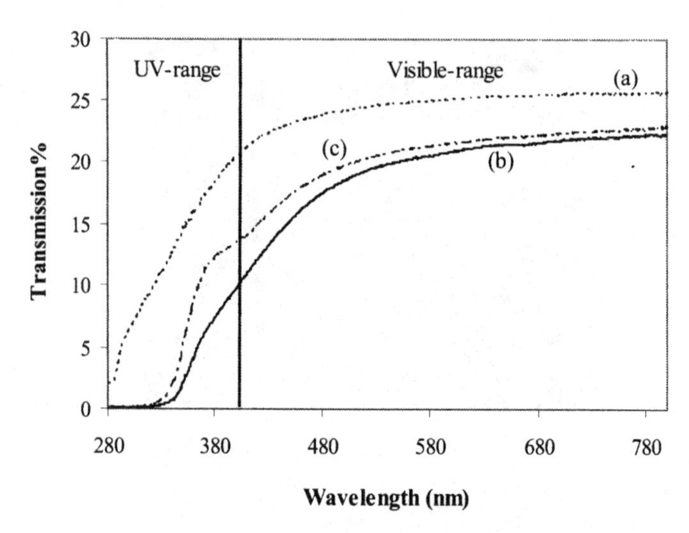

Figure 2. UV-Vis transmission spectra of (a) cotton fabric before coating, (b) coated fabric and (c) coated fabric after 20 washings.

The photocatalytic activities of the treated fabric were qualitatively assessed by an antibacterial activity test where 2.5 x 5 cm specimens of coated and uncoated cotton were placed in intimate contact with streaks of *Klebsiella Pneumoniae* gram negative bacteria organism that were made parallel and 10 mm apart on agar plates. The plates were incubated for 24 hours at 37°C under ambient cool white fluorescent light similar to normal office lighting with a UV intensity content of 4.74 $\mu W/cm^2$. The incubated plates were then examined for interruption of growth. The clear zone beneath the specimen of coated fabric reveals an almost complete killing of the seeded bacteria, whereas there was a continuing growth of the bacteria beneath the uncoated fabric specimen (Fig. 3). This is in agreement with previous studies which showed that the antibacterial effect of TiO_2 coated materials involves not only the nullification of the viability of the bacteria, but also the destruction of the bacteria cell.[19] These results clearly indicate the antibacterial capability of the coated fabric, which is considered to be due to the photocatalytic effect of the titania coating.

The effect of the titania coating on the physical and mechanical properties of cotton was tested by comparing the properties of cotton fabrics before and after coating (Table I). The bursting strength of the coated sample was found to be 9.14 kg/cm^2, compared to 8.29 kg/cm^2 for original sample. This result demonstrated that the titanium dioxide coating induced a slight enhancement to the bursting strength. On the other hand, there was less than 10% reduction in the tensile strength as a result of the enhancement of the busting strength. The resistance to air of coated fabrics was found to be less that that of original fabrics indicating that the air permeability of treated fabrics is superior to original fabrics, which is might be due to the introduction of the nanostructure surface of the coating.

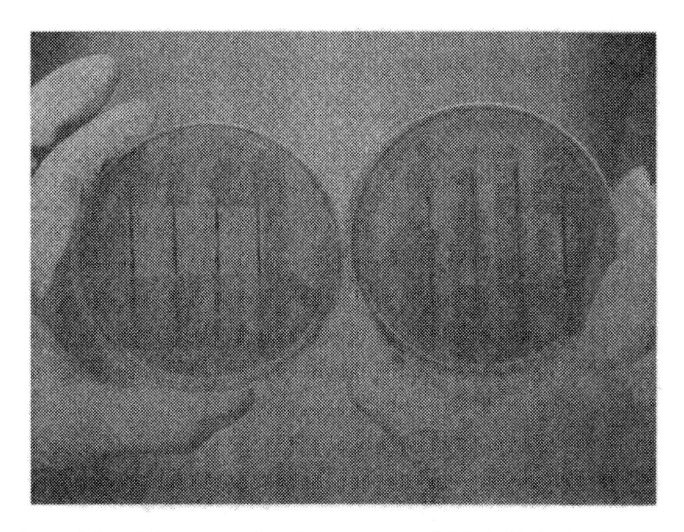

Figure 3. Antibacterial activity comparison between coated fabric (left plate) and uncoated fabric (right plate).

Table I. Mechanical properties of cotton fabrics before and after coating

		Before Coating		After Coating	
Property	Parameter (unit)	Weft*	Warp**	Weft	Warp
Tensile	Max. Load (N)	358.9	379.8	342.5	364.8
	Strain%	34.12	11.77	32.27	10.99
Bursting	Pressure (kg/cm^2)	8.29		9.14	
Permeability	R (Kpa·s/m)	0.3188		0.2978	

*Machine direction. **Cross direction.

CONSLUSIONS

A photocatalytic titania coating was produced on cotton fabrics by a low temperature sol-gel process. SEM images revealed a uniform structure of the coating films. The particles were found to be near spherical in shape with dimensions of about 20 nm. The photocatalysis of the coating was tested by means of an antibacterial activity test method. The UV absorption of the titania coating was quite substantial promoting excellent UV protection to cotton. On comparing the tensile and bursting strength before and after the treatment, it was found that the treatment had no adverse effect on the mechanical properties of cotton fabrics.

REFERENCES

[1]P. Wauthoz, M. Ruwet, T. Machej, and P. Grange, "Influence of the Preparation Method on the $V_2O_5/TiO_2/SiO_2$ Catalysts in Selective Catalytic Reduction of Nitric Oxide with Ammonia," *Appl. Catal.*, **69**, pp. 149-167, 1999.

[2]K. Kato, A. Tsuzuki, H. Taoda, Y. Torii, T. Kato, and Y. Butsugan, "Crystal Structures of TiO_2 Thin Coatings Prepared from the Alkoxide Solution via the Dip-coating Technique Affecting the Photocatalytic Decomposition of Aqueous Acetic Acid," *J. Mater. Sci.*, **29**, pp. 5911-5915, 1994.

[3]F. Victor, J. Stone, and R. J. Davis, "Synthesis, Characterization, and Photocatalytic Activity of Titania and Niobia Mesoporous Molecular Sieves," *Chem. Mater.*, **10**, pp. 1468-1474, 1998.

[4]B. O. Regan and M. Graetzel, "A Low-Cost, High-Efficiency Solar Cell Based on Dye-Sensitized Colloidal Titanium Dioxide Films," *Nature*, **353**, pp. 737-740, 1991.

[5]L. D. Birkefeld, A. M. Azad, and S. A. Akbar, "Carbon Monoxide and Hydrogen Detection by Anatase Modification of Titanium Dioxide," *J. Am. Ceram. Soc.*, **75**, pp. 2964-2968, 1992.

[6]K. Nagase, Y. Shimizu, N. Miura, and N. Yamazoe, "Preparation of Vanadium-Titanium Oxide Thin Films by Sol-Gel Processing and their Electrochromic Properties," *J. Ceram. Soc. Jpn.*, **101**, pp. 1032-1037, 1993.

[7]E. A. Barringer and H. K. Bower, "Formation, Packing, and Sintering of Monodisperse Titanium Dioxide Powders," *J. Am. Ceram. Soc.*, **65**, pp. 199-201, 1982.

[8]C. Wai, W. Y. Lin, Z. Zainal, N. E. Williams, K. Zhu, A. P. Kruzic, R. L. Smith, and K. Rajeshwar, "Bactericidal Activity of TiO_2 Photocatalyst in Aqueous Media: Toward a Solar-Assisted Water Disinfection System," *Environ. Sci. Technol.*, **28**, pp. 934-938, 1994.

[9]W. A. Badawy, "Preparation, Electrochemical, Photoelectrochemical and Solid-State Characteristics of Indium-Incorporated TiO_2 Thin Films for Solar Cell Fabrication," *J. Mater. Sci.*, **32**, pp. 4979-4984, 1997.

[10]H. Tang, K. Prasad, R. Sanjines, P. E. Schmid and F. Levy, "Electrical and Optical Properties of TiO_2 Anatase Thin Films," *J. Appl. Phys.*, **75**, pp. 2042-2047, 1994.

[11]J. P. Chatelon, A. Terrier and J. A. Royer, "Influence of Elaboration Parameters on the Properties of Tin Oxide Films Obtained by the Sol-Gel Process," *J. Sol-Gel Sci. Technol.*, **10**, pp. 55-65, 1997.

[12]D. P. Partlow, S. R. Gurkovich, K. C. Radford and L. J. Denes, "Switchable Vanadium Oxide Films by a Sol-Gel Process," *J. Appl. Phys.*, **70**, pp. 443-452, 1991.

[13]F. E. Ghodsi, F. Z. Tepehan and G. G. Tepehan, "Optical Properties of Ta_2O_5 Thin Films Deposited Using the Spin Coating Process", *Thin Solid Films*, **295**, pp. 11-15, 1997.

[14]Y. Zhang, G. Xiong, N. Yao, W. Yang, X. Fu, "Preparation of Titania-Based Catalysts for Formaldehyde Photocatalytic Oxidation from $TiCl_4$ by the Sol-Gel Method," *Catalysis Today*, **68**, pp. 89-95, 2001.

[15]U. Bach, D. Lupo, P. Comte, J. E. Moser, F. Weissortel, J. Salbeck, H. Spreitzer and M. Graetzel, "Solid-State Dye-Sensitized Mesoporous TiO_2 Solar Cells with High Photon-to-Electron Conversion Efficiencies," *Nature*, **395**, pp. 583-585, 1998.

[16]Y. Haga, H. An and R. Yosomiya, "Photoconductive Properties of TiO_2 Films Prepared by the Sol-Gel Method and its Application," *J. Mater. Sci.*, **32**, pp. 3183-3188, 1997.

[17]Technical manual of the American association of textile chemists and colorists.

[18]Annual Book of the American Society for Testing and Materials.

[19]K. Sunada, Y. Kikuchi, K. Hashimoto, and A. Fujishima, "Bactericidal and Detoxification Effects of TiO_2 Thin Film Photocatalysts," *Environ. Sci. Technol.*, **32**, pp. 726-728, 1998.

EFFECT OF SURFACE TREATMENT ON CHIRAL AND ACHIRAL $SrTiO_3$ SURFACE MORPHOLOGY AND METAL THIN FILM GROWTH

Andrew J. Francis and Paul A. Salvador
Department of Materials Science and Engineering
Carnegie Mellon University
5000 Forbes Ave.
Pittsburgh, PA 15213

ABSTRACT

Pulsed Laser Deposition has been used to grow Pt thin films on single-crystal $SrTiO_3$ substrates, both (100)-oriented (an achiral surface) and (621)-oriented (a chiral surface). Prior to growth, some of the substrates were subjected to chemical and thermal treatments. The impact of these processes on substrate surface quality, as well as the crystalline quality of the resulting Pt films, is investigated. Pt films were deposited over a range of conditions, then characterized for their crystallinity and epitaxy using x-ray diffraction (XRD) and for their surface morphologies using atomic force microscopy (AFM). For the low-index, achiral $SrTiO_3$(100) surface, pre-treatment was found to have a substantial positive effect on the substrate surface quality, leading to a well-crystallized "step-terrace" surface structure. However, similar pre-treatments did not produce a discernible step-terrace structure on $SrTiO_3$(621) surfaces, instead leading to formation of large island-like surface defects. The crystallinity and epitaxy of films deposited on pre-treated $SrTiO_3$(100) surfaces were much improved compared to films on as-received substrates. Untreated substrates led to polycrystalline, multi-oriented films while films grown on well-treated substrates were epitaxially oriented. These observations are contrasted with the successful epitaxial growth of Pt on untreated $SrTiO_3$(621) substrates. For both types of surfaces, the interface between Pt and $SrTiO_3$ will necessarily play the key role in epitaxial growth. A well-crystallized, morphologically flat substrate surface is usually considered to be a prerequisite to a low-energy metal-ceramic interface and thus high-quality films. We believe that the differences in the structural nature of the two surfaces may account for the observed differences between chiral and achiral surfaces.

INTRODUCTION

Thin metal films are interesting for use in a wide variety of applications, including electrodes in ferroelectric devices and layers in magnetic storage materials.[1] These films have been deposited on a wide variety of substrates, often taking advantage of a good structural match between film and substrate to obtain heteroepitaxially-oriented films. In particular, many studies have concerned the growth of oriented Pt films, on a number of ceramic substrates including MgO(100),[2-8] $SrTiO_3$(100),[9-11] and $SrTiO_3$(111).[12] Interest in metal-ceramic heteroepitaxy has been further increased by the recent demonstration that certain high-Miller-index surfaces ($h \neq k \neq l \neq 0$) of cubic metals exhibit enantiospecific (i.e. chiral) behavior.[13-15] One such surface, an ideal Pt(621) crystal plane, is illustrated in Fig. 1. These surfaces are currently obtained by simply cleaving metal single crystals, but to fully exploit their potential for chiral processes, methods for obtaining samples having large chiral surface areas will be necessary. Thin film deposition is naturally a good candidate for developing materials of this type.

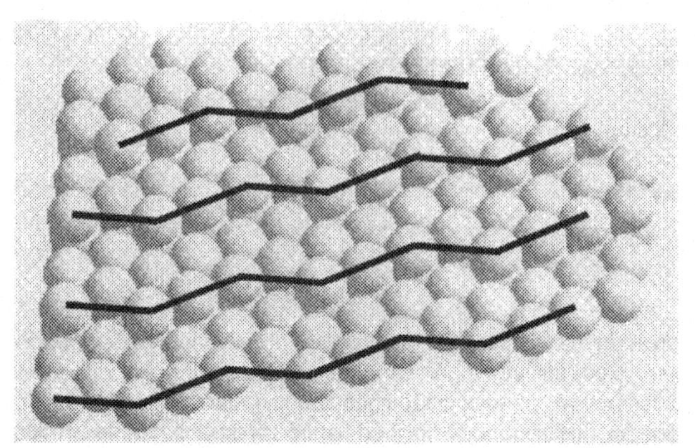

Figure 1. Representation of an ideal Pt(621) surface. Black lines indicate kinked steps on the surface. The height difference between terraces is 1.912 Å, or one half of the (100) interplanar spacing.

To encourage heteroepitaxial growth of metal films on ceramic substrates, chiral or achiral, it is helpful to have a low-energy interface between the substrate and the growing film. This is usually achieved in part by selecting substrate materials that are similar structurally to the desired film phase, to serve as templates for epitaxial growth. The focus of this paper will be the Pt/$SrTiO_3$ system. These two materials have similar crystal arrangements, as Pt is an fcc metal and $SrTiO_3$ is a perovskite, an fcc derivative structure. In addition, the Pt and $SrTiO_3$ have a lattice mismatch of just 0.5%.[10] As expected, the combination of these two materials often leads to good quality epitaxial films.

In addition to a good structural match, growth of quality films can be assisted by depositing on morphologically flat, well-crystallized substrate surfaces. Although most commercially-bought substrates probably have amorphous surfaces owing to mechano-chemical polishing treatments, recent reports have demonstrated that chemical and thermal treatments can be quite effective for the production of high-quality substrate surfaces. In particular, a chemical etch followed by a high-temperature anneal has been shown to lead to a "step-terrace" structure with wide, flat, crystalline terraces separated by short steps.[16-22] This structure is characteristic of well-developed, low-index surfaces that have no natural surface features but that need vertical steps to accommodate miscuts introduced during crystal growth. Chemical and thermal treatment methods have so far been applied to a variety of single-crystal materials, in particular perovskites such as $SrTiO_3$,[16-20] $NdGaO_3$,[20, 21] and $LaAlO_3$.[22]

Surface engineering techniques can, in fact, have a significant impact on the subsequent epitaxial growth of metal thin films, as we previously demonstrated for the deposition of Pt on $SrTiO_3$(111)[12] and will explore here for $SrTiO_3$(100). In the previous work, films were grown on both treated and untreated surfaces under otherwise identical processing conditions. The films deposited on etched and annealed surfaces were of excellent crystalline quality and exhibited a low-energy in-plane epitaxial relationship with the substrate. Films deposited on untreated $SrTiO_3$(111) surfaces, however, were either randomly oriented polycrystals or else oriented films having high-energy in-plane rotational domains that were not observed in better quality samples. Thus, surface pre-treatments can affect positively the growth of thin metal films.

In this work, we investigate the use of chemical and thermal pre-treatments to engineer substrate surfaces that will promote heteroepitaxial growth of thin metal films. First, the effects of these treatments on the well-known $SrTiO_3(100)$ surface will be presented. Next, these results will be compared to experiments performed on the high-Miller-index $SrTiO_3(621)$ surface. An ideal $SrTiO_3(621)$ surface will look quite similar to the Pt surface presented in Fig. 1 because of the structural parallels between the two materials. Growth of Pt films on both treated and untreated $SrTiO_3$ substrates is also discussed, further demonstrating the positive impact that well-developed substrate surfaces can have on the quality of deposited films.

EXPERIMENTAL

Polished single crystal substrates of (100)- and (621)-oriented $SrTiO_3$ (20 x 20 x 0.5 mm for (100), 10 x 10 x 1 mm for (621), miscut <0.5°) were obtained from Crystal GmbH (Germany). After cutting the substrates into 5 x 3 mm rectangles using a diamond-impregnated wire in a paraffin/water lubricant, the samples were ultrasonically cleaned in acetone, followed by ethanol, for 5 min each. These substrates are called "as-received" samples. Chemical etching in a 3:1 HCl:HNO_3 solution was performed on some substrates, as described previously for pre-treatments of $SrTiO_3(111)$.[23] Anneals of both (100)- and (621)-oriented $SrTiO_3$ substrates were then carried out in both air and flowing oxygen at temperatures ranging from 600-1200 °C.

PLD was carried out in a Pulsed Laser Deposition System from Neocera (Beltsville, MD). The Pt target was a commercial foil 2.5 x 2.5 cm square and 0.25 mm thick from Alfa Aesar (Ward Hill, MA). $SrTiO_3$ samples were attached to a substrate heater with conductive silver paint and heated to 600 °C under a dynamic vacuum maintained at a background pressure of $<10^{-5}$ Torr using a turbomolecular pump. The deposition atmosphere was established by lowering the turbopump speed and introducing flowing oxygen gas into the chamber to attain overall dynamic pressures of 10–100mTorr O_2.

Depositions were performed using a COMPex 201 series KrF laser (λ=248 nm, pulse duration=20 ns) from Lambda Physik (Ft. Lauderdale, Florida). The laser was operated at a rate of 3 Hz, with laser energy densities at the target from 6-8 J/cm^2. The target was rotated around its center during the deposition to keep the target surface fresh, and a short 5-minute ablation was performed just prior to deposition, with the substrate shielded, to clean the target surface. The target-to-substrate distance was maintained at ≈60 mm in all experiments. The growth rate, which has been previously determined by x-ray reflectometry to be approximately 0.19 Å/s at 8 J/cm^2,[24] is low enough such that nucleation events are strongly influenced by interactions with the substrate. After deposition, the samples were cooled in a static low vacuum ($<10^{-3}$ Torr).

The films' crystal structures were characterized by x-ray diffraction (XRD). Films on $SrTiO_3(100)$ were examined using a Rigaku diffractometer (Rigaku, Japan) equipped with Cu radiation ($K\alpha_1$ and $K\alpha_2$), and operated at 35 kV and 20 mA. 2θ–θ scans were carried out in Bragg-Brentano geometry to verify film crystallinity and orientation. ϕ–scans on (100)-oriented films, and all scans taken from Pt films on $SrTiO_3(621)$, were performed using a 3–circle Philips X'Pert system (Philips Analytical X-Ray B.V., The Netherlands). Details of this apparatus and the experimental procedure can be found in the previous report.[24]

Film surface morphologies were analyzed with atomic force microscopy (AFM) using an AutoProbe CP atomic force microscope (Park Scientific Instruments, Sunnyvale, CA), fitted with a 5μm scan head for optimal lateral resolution. Scanning was performed in contact mode using gold-coated sharpened microlever D tips from Veeco (Woodbury, NY). The AFM was operated at scan rate=2 Hz, force=1.5 pN, and gain=0.3.

RESULTS AND DISCUSSION

During the processing of single-crystalline substrate materials, the vendor subjects the wafers to a mechano-chemical polishing treatment to obtain surfaces that are as flat as possible. As illustrated in the AFM topographs in Fig. 2, as-received substrates typically have flat, morphologically featureless surfaces. The rms roughnesses for both surfaces are approximately 1.5 Å, which is very low. Fig. 2a shows that no features typically associated with a well-crystallized low-index surface[16,17] are present for as-received $SrTiO_3(100)$ surfaces. This indicates that some atomic rearrangement of the surface will be necessary before high-quality thin films can be deposited. AFM images of as-received $SrTiO_3(621)$ surfaces (Fig. 2b) also appear flat and featureless, just like images of $SrTiO_3(100)$ samples. No experimental information that correlates surface crystallography to surface morphology is available for $SrTiO_3(621)$ or other high-index surfaces. Surface features are expected to be present on an ideal high-index surface (see Fig. 1), but they are too closely spaced to be observed with the AFM technique. Additional steps introduced by crystal miscut may also be difficult to observe on this feature-dense surface, so the character of as-received $SrTiO_3(621)$ substrates cannot be directly inferred. Nonetheless, it is likely that the polishing damage leaves them poorly crystallized.

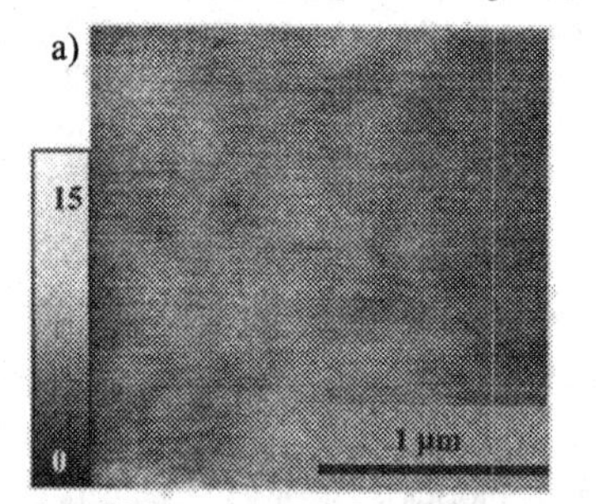

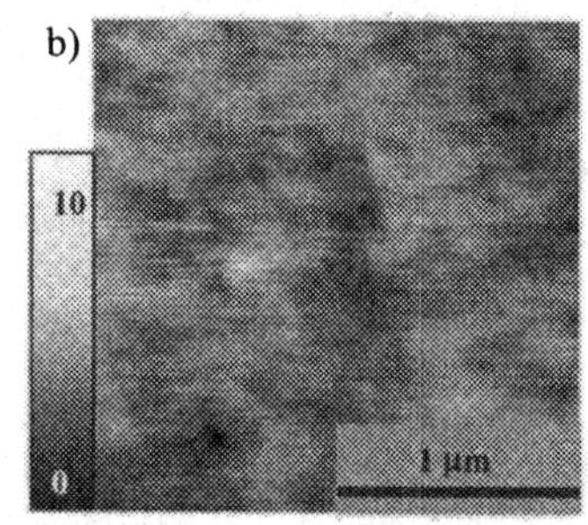

Figure 2. AFM topographic images of as-received $SrTiO_3$ a) (100) and b) (621) surfaces. The rms roughnesses of both are about 1.5Å .

Literature results suggest that the best surfaces are obtained with a short chemical etching, followed by a high-temperature annealing. The heating step is intended to recrystallize the surface by providing the surface atoms with enough energy to rearrange themselves diffusionally. This typically results in wide, flat terraces of low-index crystal planes. Therefore, from the perspective of obtaining crystalline surfaces for thin film deposition, annealing is by far the most important step and the one we will focus primarily on in this paper. Etching is primarily intended to achieve uniformly terminated surfaces. For example, an $SrTiO_3(100)$ surface can be terminated by a SrO layer or by a TiO_2 layer (as-received crystals have both terminations in roughly equal amounts), but an appropriate etchant step can result in a pure TiO_2 surface.[19, 20]

The effect of annealing is quite dramatic for $SrTiO_3(100)$ surfaces, as illustrated by the AFM image in Fig. 3. After an annealing for 2 hours in air at 1200 °C, the amorphous, featureless surface characteristic of an as-received sample has been transformed into a regular step-terrace structure, which is similar to those that have been confirmed by Koster et al. to be well-crystallized.[16, 17] Similar results were obtained over a range of temperatures from 800-1200 °C. These observations are consistent with literature results suggesting that temperatures greater than 800 °C are necessary to obtain $SrTiO_3$ surfaces having step-terrace morphologies,[10, 16, 18, 23] which will be shown to be more suitable for heteroepitaxial thin film growth than the as-received case.

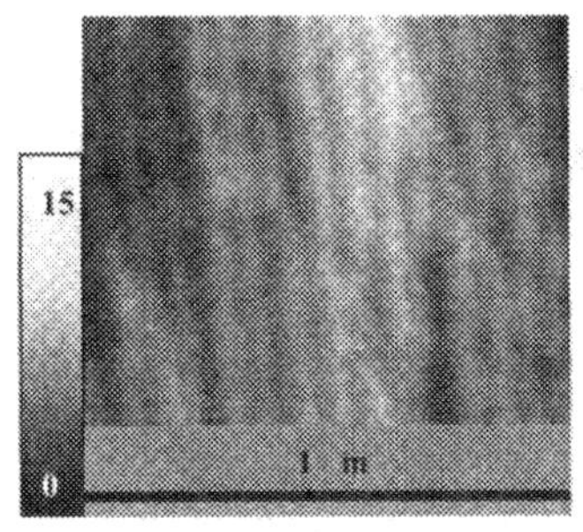

Figure 3. AFM topographic image of an $SrTiO_3(100)$ surface annealed in air for 2 hours at 1200 °C. The rms roughness is 2 Å.

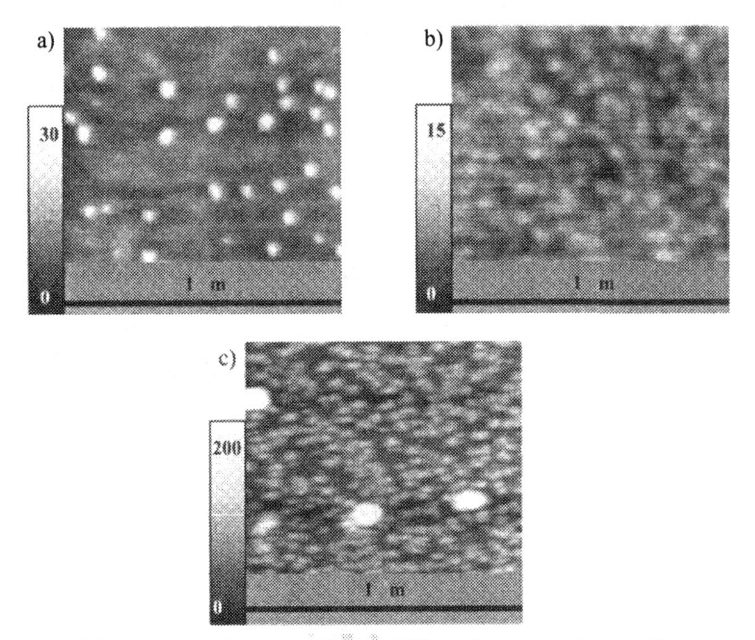

Figure 4. AFM topographic images of $SrTiO_3(621)$ surfaces annealed for 6 hours in oxygen at a) 600 °C, b) 800 °C, and c) 1200 °C. See text for rms roughness values.

These results can be compared and contrasted to those obtained after similar pre-treatments on the $SrTiO_3(621)$ surface. Fig. 4 shows several AFM images taken from $SrTiO_3(621)$ surfaces after being subjected to 6 hour thermal treatments in oxygen at various temperatures. Annealings performed at temperatures of 400 °C and below had no visible effect on the surface morphology of $SrTiO_3(621)$ substrates. Higher-temperature annealings begin to change the observed surface structures, beginning at 600 °C (Fig. 4a), where small islands start to appear on the surface. The density of these islands increases as the annealing temperature is raised to 800 °C (Fig. 4b). The rms roughnesses of these surfaces are 4 Å and 2 Å, respectively, but these values do not accurately reflect the fact that, from the perspective of thin film growth, the surface quality is deteriorating since the amount of flat substrate surface is reduced. Finally, a $SrTiO_3$(621) sample annealed at 1200 °C has an extremely rough surface relative to the as-received condition, with an rms roughness of 38 Å that does represent the poor appearance of this surface.

We believe that the differences between these two surfaces are responsible in part for the differences in their annealing behavior. For one, {621} planes are not preferred orientations of an $SrTiO_3$ crystal as it, like most other materials, prefers energetically to expose low-index facets like {100}.[25] The (621) plane lies just outside the surface stability limit observed by Sano and Rohrer at 1400 °C,[25] meaning that at very high temperatures there will be a thermodynamic instability of {621}-type surfaces. This energetic instability may represent a greater driving force for surface rearrangement that can cause large, undesirable features to develop on $SrTiO_3$(621) at the same temperatures that cause long-range ordering to take place on the stable $SrTiO_3$(100) surface. Kinetics may also play a role in the surface atoms' abilities to arrange themselves on the two types of crystals. As shown in Fig. 1, atoms need only to diffuse short distances (a few atoms) to achieve step-kink-terrace structures characteristic of well-ordered high-index surfaces. However, the steps on an ideal low-index surface may be hundreds of angstroms apart, requiring significantly longer times to form a step-terrace morphology.

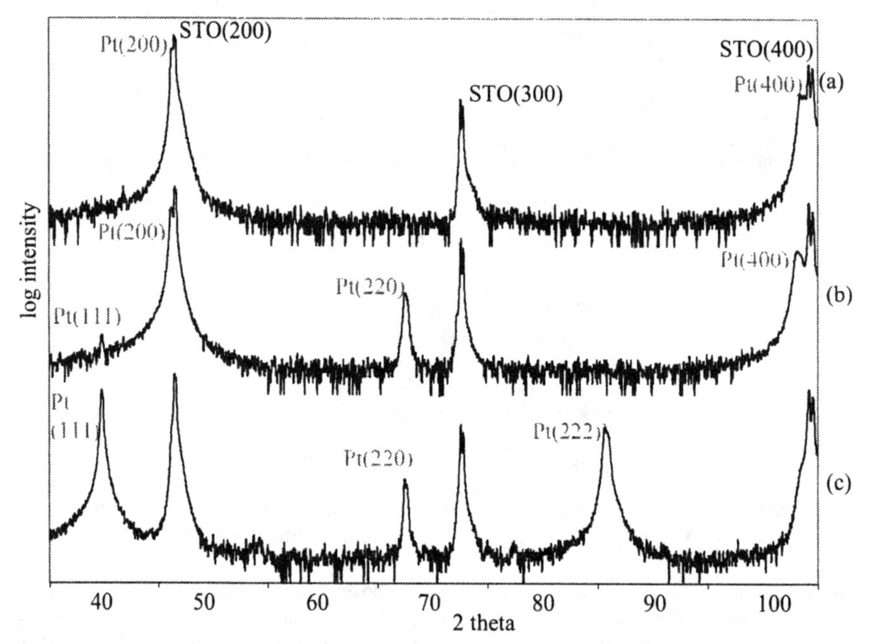

Figure 5. XRD patterns of Pt thin films deposited on $SrTiO_3$(100) surfaces that are a–b) etched and annealed and c) as-received. The films in (b) and (c) were deposited at a laser energy of 6 J/cm^2, while the film in (a) was deposited at a fluence of 8 J/cm^2. Substrate peaks are only labeled on the top figure, for clarity.

Both the quality of the substrate surface and the deposition conditions play important roles in determining the crystalline quality and epitaxial character of Pt films deposited on $SrTiO_3(100)$ substrates. X-ray diffraction results from films deposited under various conditions are presented in Figs. 5-6. Fig. 5 contains full 2θ-θ scans, while Fig. 6 shows the Pt and $SrTiO_3(200)$ peaks from each experiment in greater detail, since they are a good indicator of film quality. Epitaxial growth of Pt(100) on $SrTiO_3(100)$ is made more difficult by the fact that Pt typically prefers to grow with a (111) orientation to minimize film surface energy.[3, 5, 7] This energetic preference must therefore be overcome to realize heteroepitaxial growth of Pt(100), again underscoring the necessity of providing a low-energy interface between the two materials during growth.

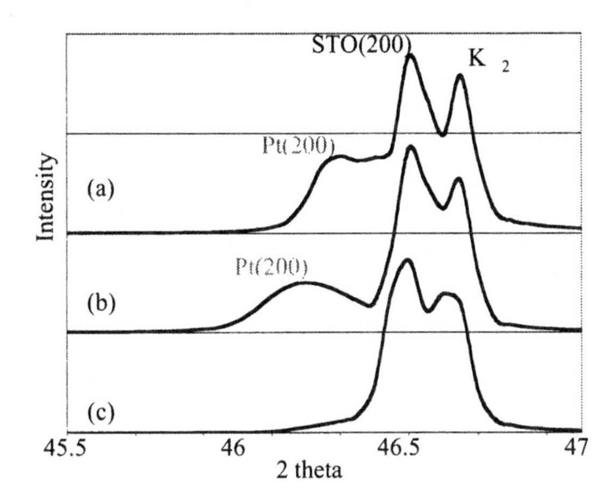

Figure 6. XRD patterns from the same samples as in Fig. 5, magnified to more clearly show the Pt and $SrTiO_3$ (200) peaks.

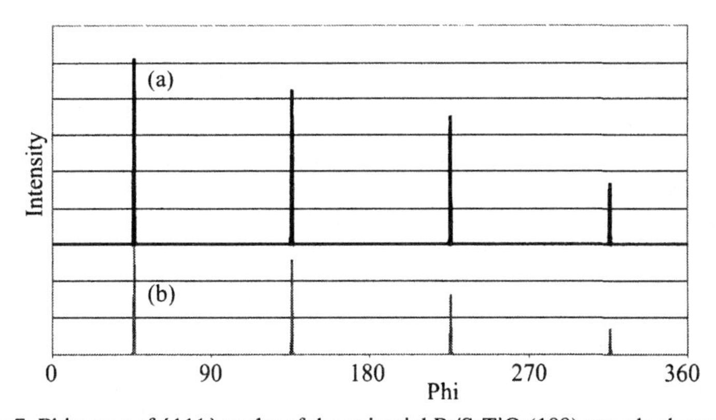

Figure 7. Phi scans of {111} peaks of the epitaxial $Pt/SrTiO_3(100)$ sample shown in Figs. 5-6a. The peaks are from a) $SrTiO_3$ and b) Pt.

To understand the effect of surface treatment on film growth, it is first necessary to determine appropriate deposition parameters, in particular the substrate temperature and laser energy density. The effect of temperature on Pt(100) growth has already been well-studied, with the consistent conclusion that temperatures greater than 500 °C are required for heteroepitaxial Pt growth.[6-8] In our work, we observed a similar critical temperature for film growth; hence, the experiments presented in this paper all involve deposition at 600 °C. The energy density of the laser beam is also an important factor, as illustrated in Figs. 5-6 (a-b) for samples deposited on etched and annealed $SrTiO_3(100)$ substrates. Increasing the laser energy density from 6 J/cm^2 (Figs. 5-6b) to 8 J/cm^2 (Figs. 5-6a) causes an improvement in Pt(100) heteroepitaxy. The percentage of crystallites having each orientation was calculated by taking the measured intensities and dividing by expected relative intensities (with multiplicity factored out). The higher laser energy boosts the percentage of epitaxially-oriented Pt calculated in this fashion from 98.4% to 100%. The Pt(200) peak height also becomes 50% more intense and much sharper for the film deposited at higher laser energy (Fig. 6a). Some of the intensity may be a function the films deposited at higher energy simply being thicker; the growth rate at 6 J/cm^2 has not been determined. Phi scans (Fig. 7) confirm "cube-on-cube" epitaxy, with the orientation relationship $(100)_{Pt}:(100)_{SrTiO_3} \parallel [010]_{Pt}:[010]_{SrTiO_3}$. We believe that a greater laser fluence has an effect similar to increased substrate temperature, where increased mobility of surface adatoms allows them to adopt a low-energy, epitaxial orientation.

The important effect of substrate pre-treatments on Pt film quality can be seen by comparing Figs. 5-6 (b and c). As described in the previous paragraph, Pt films deposited on treated $SrTiO_3$ substrates are entirely or almost entirely (100)-oriented. However, as shown in Figs. 5-6 (c), a Pt film deposited (at 6 J/cm^2) on an as-received $SrTiO_3$ substrate is polycrystalline, with a number of non-epitaxial orientations, (111) chief among them. Fig. 6c reveals that little, if any, epitaxial Pt is deposited on an as-received substrate at these conditions. This result reinforces the idea that achieving heteroepitaxial growth is contingent upon facilitating a low-energy interface between the film and the substrate, and that an as-received sample probably does not have the surface crystalline quality that might lead to such an interface. The influence of surface pre-treatment observed here is also in good agreement with that observed for Pt on $SrTiO_3(111)$ surfaces.[12]

As was the case for surface treatments, the situation is different for Pt films grown on $SrTiO_3(621)$ substrates. As described in the authors' previous work[24] and illustrated in Fig. 8, good quality heteroepitaxial Pt(621) films can be obtained by depositing on as-received substrates. Fig. 8 shows the (331) peaks for Pt and $SrTiO_3$, which serve as a good indicator of epitaxy since (621) planes cannot be directly observed by XRD.[24] The film was deposited at 600 °C with a laser energy of 6 J/cm^2. This result does not imply, however, that well-crystallized $SrTiO_3(621)$ surfaces are not important for Pt film growth, as we believe that the nature of the $SrTiO_3(621)$ surface again plays a key role. As described earlier, our Pt films are grown at 600 °C to promote the epitaxial film orientation. While relatively high temperatures are necessary for $SrTiO_3(100)$ surfaces to form step-terrace surface morphologies, we believe that simply heating $SrTiO_3(621)$ to the deposition temperature of 600 °C is sufficient to produce a similar effect on the high-index $SrTiO_3(621)$. Based on the AFM images in Fig. 4, preannealing $SrTiO_3$ samples at high temperatures would in fact produce substrate surfaces less conducive to epitaxial Pt growth. The limitations of our experimental techniques prevent us from directly determining if an $SrTiO_3(621)$ surface heated to the deposition temperature actually resembles an ideal, well-crystallized one, but the growth of Pt films on as-received substrates makes a strong case for this hypothesis.

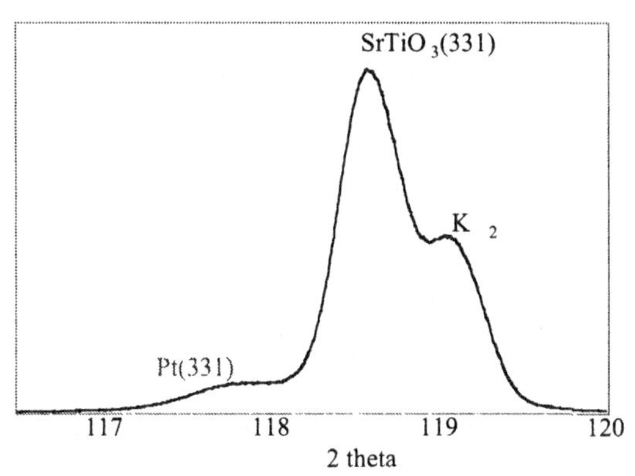

Figure 8. XRD patterns from the (331) peaks of Pt films grown on an untreated $SrTiO_3$(621) surface.

CONCLUSIONS

The effects of chemical and thermal pre-treatments on $SrTiO_3$(100) and (621) surface quality and subsequent film growth have been investigated. For the (100) surface, appropriate treatments produce a well-crystallized surface having a step-terrace structure. Films grown on these well-treated surfaces are epitaxially-oriented Pt(100), while films grown on as-received substrates are polycrystalline with a majority (111) orientation. Heat treatments on $SrTiO_3$(621) surfaces were not as effective, often producing large surface defects at high temperatures. As reported previously, epitaxial Pt(621) films can be grown on as-received $SrTiO_3$(621) substrates. These differences between the low-index and high-index surfaces are likely related to the fundamental differences in surface character between them, particularly the spacings between surface steps.

ACKNOWLEDGMENTS

The authors would like to thank Adam Dodd for assistance with the annealing studies on $SrTiO_3$(621) substrates. AJF was supported by a National Defense Science and Engineering Graduate fellowship sponsored by the Office of the Deputy Under Secretary of Defense for Science and Technology and the Army Research Office. This work was supported partially by the MRSEC program of the National Science Foundation under Award Number DMR-0079996.

REFERENCES

[1] *Handbook of Thin Film Process Technology*, D.A. Glocker and S.I. Shah ed., Institute of Physics Pub., Bristol, UK, 1995.

[2] J.F.M. Cillessen, R.M. Wolf and D.M.d. Leeuw, "Pulsed Laser Deposition of Heteroepitaxial Thin Pt Films on MgO(100)," *Thin Solid Films*, **226** 53-8 (1993).

[3] P.C. McIntyre, C.J. Maggiore and M. Nastasi, "Orientation Selection in Thin Platinum Films on (001) MgO," *J. Appl. Phys.*, **77** [12] 6201-4 (1995).

[4] P.C. McIntyre, C.J. Maggiore and M. Nastasi, "Epitaxy of Pt Thin Films on (001) MgO," *Acta Mater.*, **45** [2] 869-78 (1997).

[5] M. Morcrette, A. Gutierrez-Llorente, W. Seiler, J. Perrière, A. Laurent and P. Barboux, "Epitaxial Growth of Pt and Oxide Multilayers on MgO by Laser Ablation," *J. Appl. Phys.*, **88** [9] 5100-6 (2000).

[6] C. Gatel, P. Baules and E. Snoeck, "Morphology of Pt islands grown on MgO(001)," *J. Cryst. Growth*, **252** [1-3] 424-32 (2003).
[7] J. Narayan, P. Tiwari, K. Jagannadham and O.W. Holland, "Formation of Epitaxial and Textured Platinum Films on Ceramics-(100) MgO Single Crystals by Pulsed Laser Deposition," *Appl. Phys. Lett.*, **64** [16] 2093-5 (1993).
[8] K.H. Ahn, S. Baik and S.S. Kim, "Change of growth orientation in Pt films epitaxially grown on MgO(001) substrates by sputtering," *J. Mater. Res.*, **17** [9] 2334-8 (2002).
[9] T. Wagner, A.D. Polli, G. Richter and H. Stanzick, "Epitaxial Growth of Metals on (100) $SrTiO_3$: The Influence of Lattice Mismatch and Reactivity," *Z. Metallkd.*, **92** 701-6 (2001).
[10] A.D. Polli, T. Wagner, T. Gemming and M. Rühle, "Growth of Platinum on TiO_2-and SrO-terminated $SrTiO_3$ (100)," *Surf. Sci.*, **448** 279-89 (2000).
[11] B.S. Kwak, P.N. First, A. Erbil, B.J. Wilkens, J.D. Budai, M.F. Chisholm and L.A. Boatner, "Study of epitaxial platinum thin films grown by metalorganic chemical vapor deposition," *J. Appl. Phys.*, **72** [8] 3735-40 (1992).
[12] A. Asthagiri, C. Niederberger, A.J. Francis, L.M. Porter, P.A. Salvador and D.S. Sholl, "Thin Pt films on the polar $SrTiO_3$(111) surface: an experimental and theoretical study," *Surf. Sci.*, **537** [1-3] 134-52 (2003).
[13] G.A. Attard, "Electrochemical Studies of Enantioselectivity at Chiral Metal Surfaces," *J. Phys. Chem. B*, **105** [16] A-J (2001).
[14] J.D. Horvath and A.J. Gellman, "Enantiospecific Desorption of *R*- and *S*-Propylene Oxide from a Chiral Cu(643) Surface," *J. Am. Chem. Soc.*, **123** 7953-4 (2001).
[15] J.D. Horvath and A.J. Gellman, "Enantiospecific Desorption of Chiral Compounds from Chiral Cu(643) and Achiral Cu(111) Surfaces," *J. Am. Chem. Soc.*, **124** [10] 2384-92 (2001).
[16] G. Koster, G. Rijnders, D.H.A. Blank and H. Rogalla, "Surface Morphology Determined by (001) Single-crystal $SrTiO_3$ Termination," *Physica C*, **339** 215-30 (2000).
[17] G. Koster, B.L. Kropman, G.J.H.M. Rjinders, D.H.A. Blank and H. Rogalla, "Quasi-ideal strontium titanate crystal surfaces through formation of strontium hydroxide," *Appl. Phys. Lett.*, **73** [20] 2920-2 (1998).
[18] H. Bando, Y. Aiura, Y. Haruyama, T. Shimizu and Y. Nishihara, "Structure and electronic states on reduced $SrTiO_3$(100) surface observed by scanning tunneling microscopy and spectroscopy," *J. Vac. Sci. Technol. B*, **13** [3] 1150-4 (1995).
[19] M. Kawasaki, K. Takahashi, T. Maeda, R. Tsuchiya, M. Shinohara, O. Ishiyama, T. Yonezawa, M. Yoshimoto and H. Koinuma, "Atomic Control of the $SrTiO_3$ Crystal Surface," *Science*, **266** 1540-2 (1994).
[20] V. Leca, G. Rijnders, G. Koster, D.H.A. Blank and H. Rogalla, Mat. Res. Symp. Soc. Proc. **587,** O3.6.1-O3.6.4 (2000).
[21] T. Ohnishi, K. Takahashi, M. Nakamura, M. Kawasaki, M. Yoshimoto and H. Koinuma, "A-site layer terminated perovskite substrate: $NdGaO_3$," *Appl. Phys. Lett.*, **74** [17] 2531-3 (1999).
[22] J. Yao, P.B. Merrill, S.S. Perry, D. Marton and J.W. Rabalais, "Thermal stimulation of the surface termination of $LaAlO_3$," *J. Chem. Phys,*, **108** [4] 1645-52 (1998).
[23] T.-D. Doan, J.L. Giocondi, G.S. Rohrer and P.A. Salvador, "Surface Engineering along the Close-packed Direction of $SrTiO_3$," *J. Cryst. Growth*, **225** 178-82 (2001).
[24] A.J. Francis and P.A. Salvador, "Chirally-oriented heteroepitaxial thin films grown by pulsed laser deposition: Pt(621) on $SrTiO_3$(621)," *Submitted to J. Appl. Phys.*, (2004).
[25] T. Sano, D.M. Saylor and G.S. Rohrer, "Surface Energy Anisotropy of $SrTiO_3$ at 1400 °C in Air," *J. Am. Ceram. Soc.*, **86** [11] 1933-9 (2003).

SURFACE CHARACTERIZATION OF LOW-TEMPERATURE PROCESSED TITANIA COATINGS PRODUCED ON COTTON FABRICS

Walid A. Daoud* and John H. Xin
Nanotechnology Centre,
Institute of Textiles & Clothing,
The Hong Kong Polytechnic University,
Hung Hom, Hong Kong

ABSTRACT

Thin coatings of transparent titanium dioxide were produced on cotton fabrics from alkoxide solutions at low temperatures. The structure and morphology of these coatings were investigated by scanning electron microscopy (SEM), X-ray diffraction spectroscopy (XRD), Raman spectroscopy and transmission electron microscopy (TEM). SEM images showed the formation of a continuous layer of sol-gel derived titanium dioxide with grains of about 15 nm in size. XRD, Raman spectra and TEM micrographs indicated the existence of small amount of anatase crystallites within the films.

INTRODUCTION

Crystalline titanium dioxide has received increasing attention due to their interesting properties and potential applications, e.g. photocatalysts,[1-3] photoelectrodes,[4] gas sensors,[5] and electrochromic display devices.[6,7] Nanosized TiO_2 particles show high photocatalytic activities because they have a large surface area per unit mass and volume and hence facilitate the diffusion of excited electrons and holes towards the surface before their recombination.[8] Recently, the sol-gel process and dip-coating technique were employed to produce porous anatase titania thin films.[9-11] However, high temperatures of 500-550 °C were required to produce those films.

The formation of photocatalytic titania films at low temperature is important for the fabrication of transparent films on substrates that can not withstand high temperature treatment, such as plastics and textiles. High temperature treatment would limit the utilization of the optical and photocatalytic properties of sol-gel derived titania in such application areas.

It is well-known that titania film properties are highly dependant on the preparation process and surface microstructure. Therefore, the characterization of the microstructure of titania thin films is of high significance.

In this article, we report our observations form the surface characterization of titania thin films produced by a low temperature sol-gel process and coated on cotton. On analyzing the microstructure and crystallinity of the thin films by SEM, XRD, Raman and TEM, it was found that the prepared films were of continuous and amorphous structure, however anatase crystallites were identified in a small scale within the films.

---

* Corresponding author, tcdaoud@polyu.edu.hk

EXPERIMENTAL

The nanosol was prepared at room temperature by mixing titanium tetraisopropoxide (Aldrich, 97%) with absolute ethanol (Riedel, 99.8%) at pH of 1 to 2. The mixture was vigorously stirred for one hour prior to coating.

A 10 x 10 cm knit cotton substrate was dried at 100 °C for 30 minutes, dipped in the nanosol for 30 seconds and then padded using an automatic padder at a nip pressure of 2.75 kg/cm^2. The padded substrates were then dried at 80 °C for 10 minutes in a preheated oven to drive off ethanol and finally cured at 150 °C for 5 minutes in a preheated curing oven.

The structure and morphology of these coatings were investigated using scanning electron microscopy (SEM) (Leica Stereoscan 440) equipped with (Oxford Energy Dispersive X-ray System) operating at 20 kV.

The crystallinity of the coatings were studied by X-ray diffraction spectroscopy (XRD) (Bruker D8 Discover X-ray Diffractometer) operating at 40 kV, Raman spectroscopy using a 514.5 nm laser line from a continuous wave argon laser (Coherent Innva 70) with 250 mW power, a double grating monochromator (Spex 1403) equipped with a cooled photomultiplier tube (PMT, Hamama-Tus R943-2) and high resolution transmission electron microscopy (HRTEM) (JEOL JSM-2010 microscope) operated at 200 keV. To prepare the TEM samples, titania films were produced by spin coating (1500 rpm for 40 seconds) on a silicon wafer and heat-treated in a similar manner as explained earlier for coating on cotton. The sample was then glued on a copper grid followed by mechanical grinding and polishing until the thickness was reduced to 20 μm thick. Ion milling was performed on the substrate only by Ar^+ bombardment at 5 keV using a Gatan Precision Polishing System. Initially, the angle of ion milling was set at 10 degrees; at the final stage, an angle of 7 degrees was used and processing was performed at 3 keV.

Figure 1. High magnification SEM showing the shape and average size of the particles of the titania coating.

RESULTS AND DISCUSSION

The observation of the titania films by SEM shows that the surface structure was non-porous with no grains. This is in agreement with previously reported SEM images of titania films cured by heating at 500°C in oxygen atmosphere.[12] Figure 1 is a high magnification SEM image shows near spherical shape and the 20 nm average size of the particles.

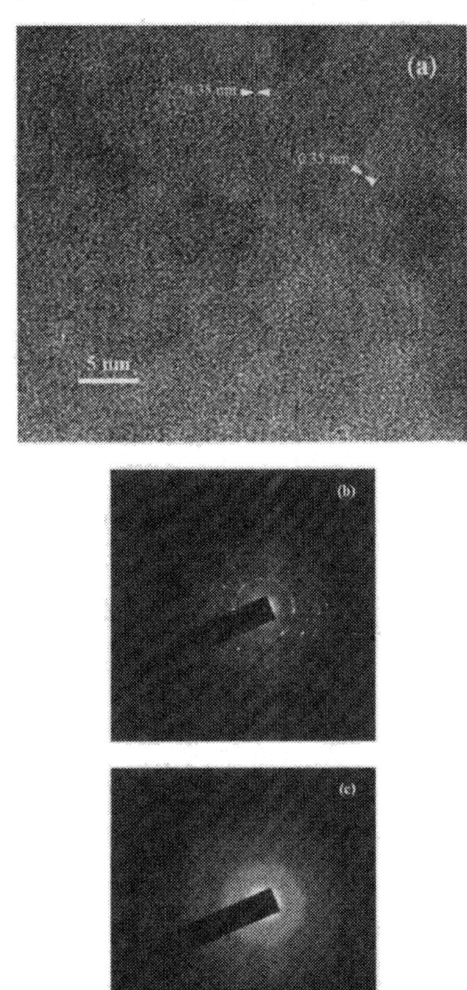

Figure 2. (a) HRTEM image of a plane view of titania film coated on a silicon wafer, (b) the corresponding SAED pattern and (c) a typical SAED pattern of predominantly amorphous regions.

Figures 2(a) and (b) show an HRTEM image and the corresponding selected area electron diffraction (SAED) pattern, respectively. Figure 2(a) shows the existence of small crystallites with a lattice fringe of 0.35 nm that corresponds to the 101 lattice plane and a diameter of 5-10 nm. Measurements of lattice spacing from the corresponding SAED pattern as shown in Figure 2(b) indicate that the nanocrystals were anatase. We also investigated other regions of the sample, where no lattice fringes could be found. A typical diffraction pattern of such regions is shown in Figure 2(c) suggesting that the film is mainly of amorphous nature. However, the peak of diffuse Debye-Scherrer rings that correspond to the lattice spacing for anatase crystallites can still be observed in such SAED patterns. Thus, it can be included that the nucleation and growth of anatase nanocrystals were at an early stage.

The crystalline and microstructure of titania films coated on cotton fabrics were studied by X-ray diffraction (XRD). The bulk of the X-ray signal originated from cotton as cotton is the underlying substrate. However, diffraction peaks associated with anatase phase at 25 and 37° were observed in a small magnitude as shown in Fig. 3. An enhancement of the cotton associated peak at 20° was also observed. The small peaks associated with crystalline titania suggest that the film is predominantly amorphous with a small content of anatase crystallites. This is well in agreement with the HRTEM observations.

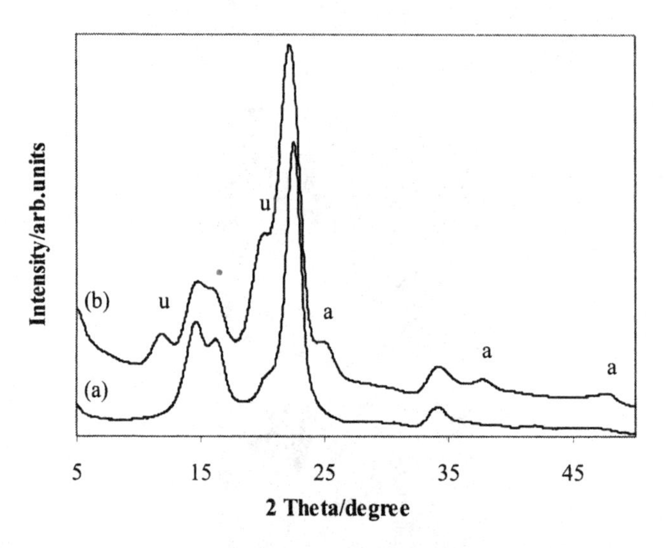

Figure 3. XRD patterns for (a) pure cotton and (b) TiO_2 film coated on cotton (a, anatase; u, unknown).

In Fig. 4 the Raman spectrum of the as-grown titania films shows peaks associated with the underlying cotton substrates as well as a broad peak at 137 – 161 cm^{-1}. This suggests the existence of small anatase crystallites embedded in amorphous regions that give rise to a broad anatase associated peak. Thus, the Raman spectra are in full agreement with the XRD and TEM observations.

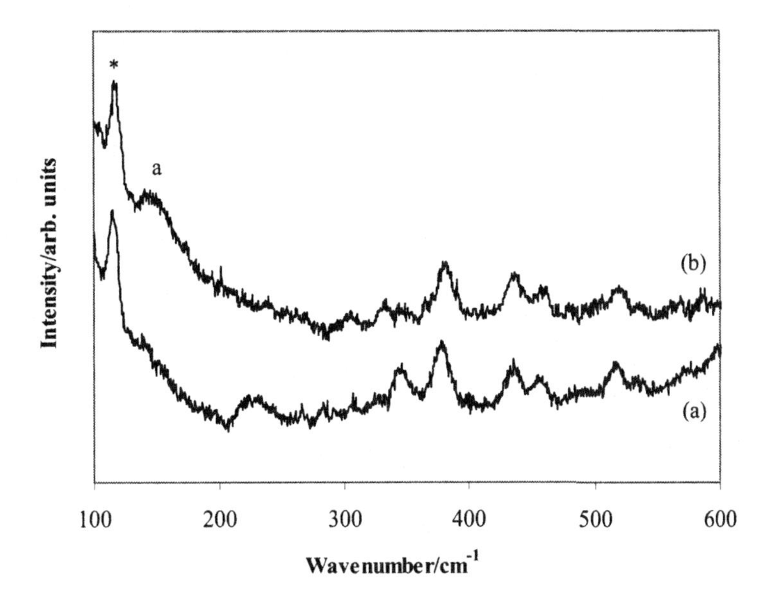

Figure 4. Raman spectra for (a) pure cotton and (b) TiO_2 film coated on cotton (a, anatase; *, laser line).

CONSLUSIONS

Nano-scaled titania layers were produced on cotton fabric by a low temperature sol-gel process. SEM shows that the particles are 20 nm in size have a near spherical shape. HRTEM micrographs together with the XRD and Raman spectra reveal the amorphous structure of the coating with a strong evidence of the existence of anatase crystallites in a small scale within the titania films.

REFERENCES

[1]P. Wauthoz, M. Ruwet, T. Machej, and P. Grange, "Influence of the Preparation Method on the $V_2O_5/TiO_2/SiO_2$ Catalysts in Selective Catalytic Reduction of Nitric Oxide with Ammonia," *Appl. Catal.*, **69**, pp. 149-167, 1999.

[2]K. Kato, A. Tsuzuki, H. Taoda, Y. Torii, T. Kato, and Y. Butsugan, "Crystal Structures of TiO_2 Thin Coatings Prepared from the Alkoxide Solution via the Dip-coating Technique Affecting the Photocatalytic Decomposition of Aqueous Acetic Acid," *J. Mater. Sci.*, **29**, pp. 5911-5915, 1994.

[3]F. Victor, J. Stone, and R. J. Davis, "Synthesis, Characterization, and Photocatalytic Activity of Titania and Niobia Mesoporous Molecular Sieves," *Chem. Mater.*, **10**, pp. 1468-1474, 1998.

[4]B. O. Regan and M. Graetzel, "A Low-Cost, High-Efficiency Solar Cell Based on Dye-Sensitized Colloidal Titanium Dioxide Films," *Nature*, **353**, pp. 737-740, 1991.

[5]L. D. Birkefeld, A. M. Azad, and S. A. Akbar, "Carbon Monoxide and Hydrogen Detection by Anatase Modification of Titanium Dioxide," *J. Am. Ceram. Soc.*, **75**, pp. 2964-2968, 1992.

[6]K. Nagase, Y. Shimizu, N. Miura, and N. Yamazoe, "Preparation of Vanadium-Titanium Oxide Thin Films by Sol-Gel Processing and their Electrochromic Properties," *J. Ceram. Soc. Jpn.*, **101**, pp. 1032-1037, 1993.

[7]E. A. Barringer and H. K. Bower, "Formation, Packing, and Sintering of Monodisperse Titanium Dioxide Powders," *J. Am. Ceram. Soc.*, **65**, pp. 199-201, 1982.

[8]M. Anpo, T, Shima, S. Kodma, Y. Kubokawa, "Photocatalytic Hydrogenation of Propyne with Water on Small-Particle Titania: Size Quantization Effects and Reaction Intermediates," *J. Phys. Chem.*, **91**, pp. 4305-4310, 1987.

[9]J. P. Chatelon, A. Terrier and J. A. Royer, "Influence of Elaboration Parameters on the Properties of Tin Oxide Films Obtained by the Sol-Gel Process," *J. Sol-Gel Sci. Technol.*, **10**, pp. 55-65, 1997.

[10]D. P. Partlow, S. R. Gurkovich, K. C. Radford and L. J. Denes, "Switchable Vanadium Oxide Films by a Sol-Gel Process," *J. Appl. Phys.*, **70**, pp. 443-452, 1991.

[11]F. E. Ghodsi, F. Z. Tepehan and G. G. Tepehan, "Optical Properties of Ta_2O_5 Thin Films Deposited Using the Spin Coating Process", *Thin Solid Films*, **295**, pp. 11-15, 1997.

[12]V. Balek, T. Mitsuhashi, I.M. Bountseva, I.N. Beckman, Z. Malek, J. Subrt, "Effect of Gas Environment on Titania Films Microstructure Characterized by Emanation Thermal Analysis," *J. Therm. Anal. Cal.*, **69**, pp. 93-101, 2002.

THERMODYNAMICS OF REFRACTORIES FOR BLACK LIQUOR GASIFICATION

Alireza Rezaie, William L. Headrick and William G. Fahrenholtz
University of Missouri-Rolla
Department of Ceramic Engineering
1870 Miner Circle
Rolla, MO 65409
USA
ardh9@umr.edu

"This report was prepared with the support of the U. S. Department of Energy, under Award No. DE-FC26-02NT41491. However, any opinions, findings, conclusions, or recommendations expressed herein are those of the authors and do not necessarily reflect the views of the DOE."

ABSTRACT

FactSage® thermodynamic software was used to analyze the phases present in black liquor and to predict the interaction of black liquor with different refractory compounds. Black liquor is a water solution of the non-cellulose portion of wood (mainly lignin) and spent pulping chemicals (Na_2CO_3, K_2CO_3, and Na_2S). Modeling included prediction of phases formed under operating conditions of a high temperature black liquor gasification (BLG) process used in the pulp and paper industry. At the operating temperature of the BLG, it was predicted that the water would evaporate from the black liquor and that the organic portion of black liquor would combust, leaving a black liquor smelt composed of sodium carbonate (70-75%), potassium carbonate (2-5%), and sodium sulfide (20-25%). Exposure of aluminosilicates to this smelt leads to significant corrosion due to formation of expansive phases and, subsequently, cracking and spalling. Oxides such as ZrO_2, CeO_2, La_2O_3, Y_2O_3, Li_2O, MgO and CaO are resistant to black liquor smelt but non-oxides such as SiC and Si_3N_4 are oxidized and dissolved by the smelt. Other candidates such as $MgAl_2O_4$ and $BaAl_2O_4$ are resistant to sodium carbonate but not to potassium carbonate. $LiAlO_2$ shows stability against both sodium carbonate and potassium carbonate. Candidate materials selected on the basis of the thermodynamic calculations are being tested by sessile drop test for corrosion resistance to molten black liquor smelt.

1. INTRODUCTION

Development of new refractory materials is a critical issue for implementation of BLG technology. Black liquor is a by-product of the papermaking process. Black liquor is an aqueous solution containing waste organic material, which is mainly lignin, as well as the spent pulping chemicals, which are primarily sodium carbonate and sodium sulfide [1]. Chemical energy can be recovered from black liquor by burning it as a liquid fuel in a boiler or gasifier. Based on its energy content, black liquor is expected to become an increasingly important resource for pulp and paper producers for electric power generation in coming years [1, 2]. Recovery boilers have been used successfully for many years, but they have a number of shortcomings including high capital expense, low efficiency, and the potential for explosion [1, 3-4].

Black Liquor Gasification is widely viewed as the technology that will replace the recovery boiler in the pulp and paper industry. Similar gasification processes are used to convert low-cost solids such as biomass or waste liquids into clean-burning gases [5]. Combustion of these gases has the potential to partially or fully meet the energy needs for pulp and paper plants, reducing or eliminating dependence on electricity generated commercially by the combustion of fossil fuels. The fundamentals of the gasification process have been reviewed elsewhere [6]. Several distinct BLG processes have been proposed, but only two of them have had satisfactory results in plant trials [7]. The first process is the low temperature process (600-700°C) developed by Manufacturing Technology Conversion International, Inc (MTCI) [8]. The other is the high temperature process (900-1000°C) developed by Chemrec [8]. The schematic configuration of a high temperature BLG reformer run under atmospheric pressure is presented in Figure 1. Both the high temperature and low temperature processes currently use air for combustion, but when oxygen is used in place of air, substantially higher gasification temperatures, up to 1400°C are possible. Higher temperatures result in higher overall process efficiencies [9].

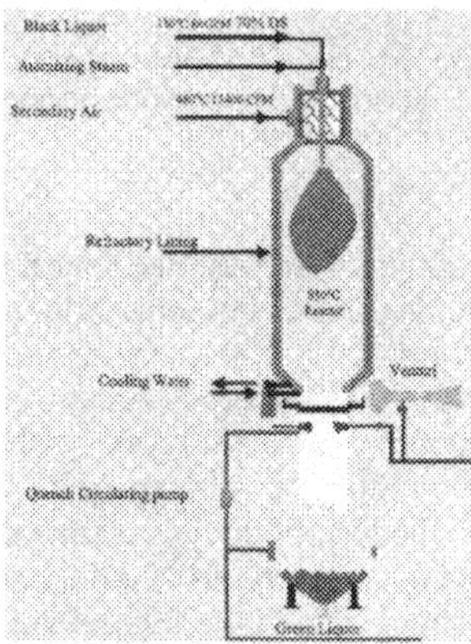

Fig. 1: High Temperature Low Pressure Black Liquor Gasification Refiner [1]

The commercial success of BLG technology requires the development of improved refractory materials for the protective lining of the gasifier. To date, aluminosilicate or fused cast alumina-based materials have been used in this application. Both thermodynamic calculations and experience show that these aluminosilicates are not sufficiently resistant to the alkali containing atmospheres for extended operation of gasifiers. The overall objective of research in progress at University of Missouri-Rolla is the development of cost-effective materials with improved performance in gasifier environments. The thermodynamics analysis reported in this paper is the first step in the development process.

2. INTERACTION OF CANDIDATE REFRACTORIES WITH BLACK LIQUOR SMELT

FactSage® 5.1 is a thermodynamic modeling package that contains a database of thermodynamic properties. It performs thermodynamic equilibrium calculation based on Gibbs free energy minimization. As with all thermodynamic models, this one predicts equilibrium and does not take into account kinetic or microstructural factors. Experiments are necessary to verify the predictions.

2.1 Corroding Smelt

The typical composition (wt %) of black liquor introduced into the BLG reformer has been reported as 35% C, 35.4% O_2, 19.4% Na, 4.2% S and 3.5% H_2 [10]. Using thermodynamic modeling, this elemental analysis can be converted into an equilibrium compound composition (Table I). The phase/compound composition was modeled at 950°C and a pressure of 1atm.

Table I. Compound composition of black liquor at T=950°C

Constituents	Na_2CO_3	Na_2S	K_2CO_3	C
Wt %	50-55	25-30	1-3	15-20

After the free carbon in black liquor is combusted, the composition of the resulting smelt that would then contact the refractory lining is 70-75% Na_2CO_3 (T_m=858°C), 20-25% Na_2S (T_m=1172°C) and 2-5% K_2CO_3 (T_m=901°C). Formation of sodium sulfate (Na_2SO_4) was predicted when the amount of oxygen introduced to the gasifier was more than the stoichiometric amount necessary to burn the organic portion of black liquor. But no data is reported about the oxygen content, in the atmosphere of existing BLG reformers.

From the prediction, about three quarters of the black liquor smelt was composed of Na_2CO_3 and K_2CO_3 forming a liquid solution at the operating temperature of high temperature black liquor gasifiers. About one quarter of the smelt was Na_2S, which should not be as corrosive as the two other components because it is in the solid state at the operating temperature of the black liquor gasifier. Consequently, the selection of refractory materials for this application must be based upon resistance to molten Na_2CO_3; although interactions with Na_2S and K_2CO_3 should not be ignored. No evidence of solubility of Na_2S in liquid was found in available phase diagrams or predicted by FactSage®. Na_2SO_4 melts at 884°C, which was below the operating temperature of BLG; therefore, if sodium sulfate forms, it should be considered as a liquid part of the smelt in contact with refractory. The analysis that follows has considered only the presence of Na_2S.

2.2 Interaction of Aluminosilicates with Black Liquor

The main compounds predicted to be present in aluminosilicate refractories were corundum, meta-stable aluminosilicate compounds (andalusite, silimanite or kyanite) and mullite. Because the aluminosilicate phases are meta-stable at 950°C, only the stable compounds (e.g., mullite and alumina) were modeled in this analysis. If these refractory compounds were exposed to black liquor smelt at 950°C and total pressure (P_t) of 1atm, the reaction products for corundum would be mainly β"-alumina ($Na_2O•12Al_2O_3$) (~75%) and β-alumina (10%). For mullite, the products would be nepheline (~50%) and corundum (~40%). The total composition is summarized in Table II. The atmosphere inside the BLG gasifier formed under these conditions would be mainly composed of CO and H_2.

Table II. Products of the reaction between black liquor smelt and aluminosilicate refractories at 950°C

Reaction Products / Refractory Compound	Corundum	β"-alumina	β-alumina	K-β"-alumina	Nepheline	Albite	Leucite
Corundum		×	×	×			
Mullite	×				×	×	×

(×): the phase is formed

To summarize the information presented in Table II, it was predicted that none of the aluminosilicate refractory compounds would be resistant to either Na_2CO_3 or black liquor smelt at 950°C, which is the temperature of gasification in high temperature processes. Products formed by the reaction of black liquor smelt with corundum have a substantially larger volume than the original refractory. For example, mullite exposed to black liquor at 1000°C, showed about 30% volume expansion upon reaction. Alumina refractories showed 13% expansion after reaction. Only 0.7% expansion is reported for α-alumina + β-alumina under the same conditions [11]. In use, the volume expansion promotes crack formation and spallation of the refractory, which then exposes the underlying materials allowing for further attack.

In the case of mullite, thermodynamic equilibrium calculations predicted that nepheline, albite, leucite and corundum would form. All of these phases were in the solid state under BLG operating conditions. By the same mechanism as described for corundum, large volume increases accompanied by formation of new phases such as nepheline (about 30%) would be enough to nucleate and propagate cracks in a structure, decreasing the lining life due to spallation. Corundum formed as the reaction product of mullite with black liquor would be attacked again with black liquor smelt and corroded by the same mechanism. The vaporization of refractory constituents was negligible under these conditions.

2.3 Interaction of Simple Oxide and Non-Oxide Refractories with Black Liquor

Thermodynamic studies were performed to predict the behavior of some simple refractory oxides and complex oxides, as well as non-oxides as new refractory materials, against Na_2O, Na_2CO_3, K_2O, and K_2CO_3. Simple oxides selected as candidates for use in high temperature black liquor gasifier are Al_2O_3, SiO_2, MgO, CaO, ZrO_2, Y_2O_3, La_2O_3, CeO_2, Li_2O and BaO. As the first step, an effort was made to plot an Ellingham Diagrams (Figure 2) to see the potential of sodium or potassium metal vapor to reduce the oxide candidates because the existence of alkaline metal vapor in the gasifier atmosphere was possible. If the free energy of formation of each candidate was less than that for sodium or potassium oxide, that oxide was more stable than sodium (potassium) oxide and sodium (potassium) metal vapor was not able to reduce it. The total pressure (P_t) selected to plot the diagram is 1 atmosphere. Based on the diagram, all of candidate simple oxides were resistant to sodium or potassium metal vapor at P_t=1atm and T=300-1100°C because they all had free Gibbs energy of formation less than that of sodium or potassium oxide. However, this criterion was not sufficient to evaluate the materials due to lack of information about new compounds that may be formed. For example, alumina or silica was not reduced by sodium, but they were known to form new compounds which may lead to failure.

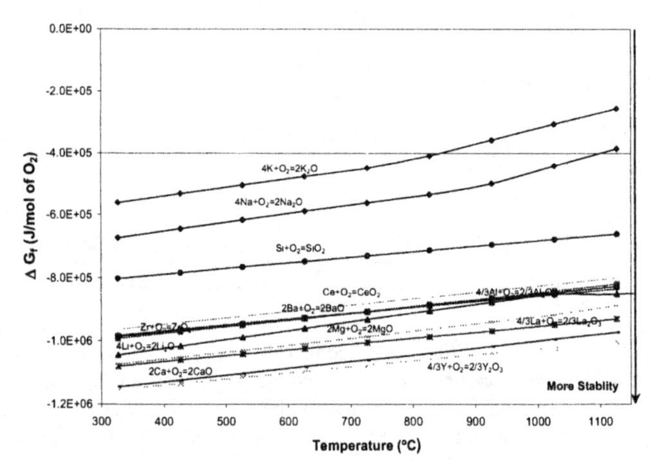

Fig. 2: Ellingham Diagram of candidate simple oxides against sodium

Thermodynamic modeling was also used to predict the behavior of candidate refractory simple oxides against the main components of black liquor at T=900-1000°C. The results are listed in Table III which show that all candidates except Al_2O_3, SiO_2 and BaO were resistant against black liquor. Moreover, it was observed that SiC and Si_3N_4, two non-oxide refractory candidates for black liquor gasifier applications, were not resistant to black liquor constituents. SiC was converted to compounds such as $(Na_2O)(SiO_2)$, $Na_6Si_2O_7$, K_2SiO_3 and $K_2Si_2O_5$; where some of them were in liquid state at operating temperature of black liquor gasifier and would dissolve into the smelt.

Table III. Interaction of refractory simple oxides and non-oxide with black liquor components at T=900-1000°C

Refractory	Al_2O_3	SiO_2	BaO	MgO	CaO	ZrO_2	Y_2O_3	La_2O_3	CeO_2	Li_2O	SiC	Si_3N_4
Na_2O	×	×	×								×	×
Na_2CO_3	×	×	×								×	×
K_2O	×	×	×								×	×
K_2CO_3	×	×	×								×	×

(×): reaction

Despite their resistance to direct reaction with black liquor, hydration of MgO and CaO was a concern especially when the operating conditions of the gasifier included water vapor. An effort was made to predict the hydration behavior of these oxides as a function of temperature at P_{H2O} = 1atm (Figure 3).

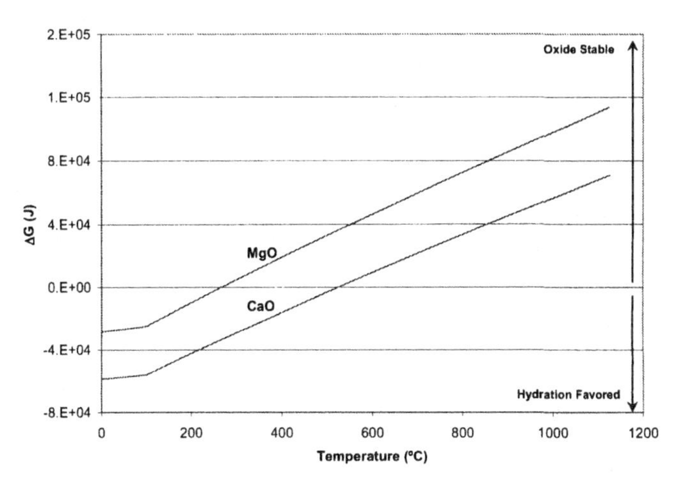

Fig. 3: Hydration behavior of Magnesia and calcia

In use, a temperature gradient exists across refractory linings. The hydration of CaO and MgO depends on penetration of water vapor into the refractory lining through porosity, cracks or joints. Figure 3 shows that at P_{H2O}= 1atm, magnesia hydrates only below 266°C. Calcia hydrates easily at temperatures up to 524°C. Therefore, magnesia is a better choice for application in black liquor gasifier because of its higher resistance to hydration.

2.4 Interaction of Aluminates with Black Liquor

Aluminates are also candidates for BLG applications. Thus, the behavior of aluminates in alkaline-containing atmospheres must be investigated. The aluminates considered as part of this investigation were $MgAl_2O_4$, $BaAl_2O_4$ and $LiAlO_2$. As a first step, the interaction of aluminates with Na_2O was considered. Na_2O can form under non-steady state operations of the BLG. Equation 1 describes the main corrosion reaction of these oxides with sodium oxide:

$$^{*}M_nAl_2O_4 + Na_2O \longrightarrow M_nO + 2NaAlO_2 \quad (1)$$

The change in the free Gibbs Energy (ΔG) of reaction versus temperature is shown in Figure 4 for each of the aluminates. None of the aluminates are resistant to sodium oxide up to 1100°C because ΔG of reaction with sodium oxide for all of them was negative. It could be concluded that barium aluminate was the most and magnesium aluminate was the least resistant. Because sodium was in the form of sodium carbonate in black liquor and in the normal working conditions of high temperature gasifiers, prediction of the behavior of these aluminates to sodium carbonate was important.

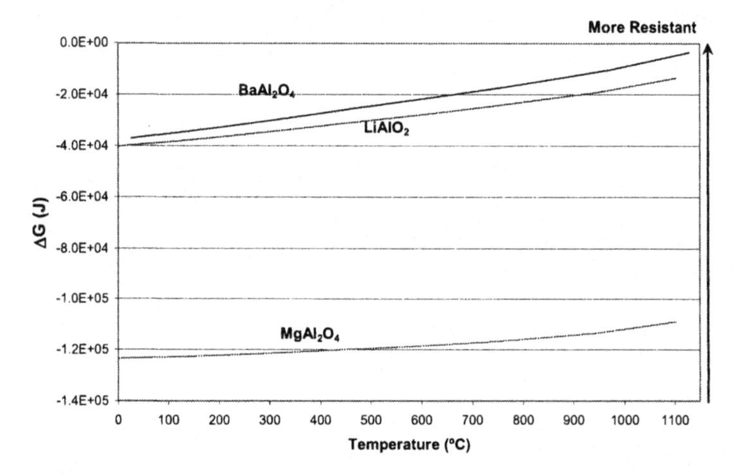

Fig. 4: ΔG for the reaction between three candidate aluminates with sodium oxide

The reaction equation of candidate aluminates with sodium carbonate was as follows:

$$M_nAl_2O_4 + Na_2CO_3 \longrightarrow M_nCO_3 + 2NaAlO_2 \quad (2)$$

---

* M: Mg, Ba, Li
$n_{(Mg, Ba)}=2$, $n_{(Li)}=1$

The range of temperature over which the carbonates are stable is important to identify the stable compounds in black liquor. The reaction equation for stability of the considered carbonates was as follows:

$$^{\dagger}MCO_3 \longrightarrow MO + CO_2 \tag{3}$$

ΔG for dissociation of carbonates as a function of temperature in the range of 300-1100°C based on the thermodynamic analysis base is presented in Figure 5.

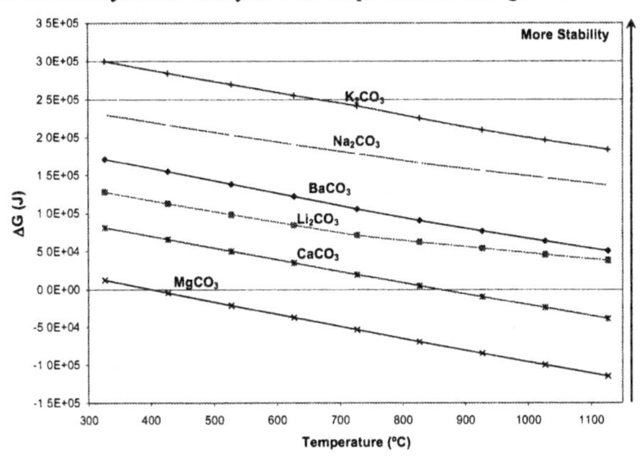

Fig. 5: ΔG of dissociation of carbonates versus temperature

All of the considered carbonates that formed as reaction products of aluminates with sodium carbonate were stable except for magnesium carbonate which would dissociate at 400°C under P_t = 1atm. Calcium carbonate also dissociated but at higher temperature, 850°C. Therefore from T= 400°C, the equation for the reaction of magnesium aluminate with sodium carbonate should change to the reaction as follows:

$$MgAl_2O_4 + Na_2CO_3 \longrightarrow 2NaAlO_2 + MgO + CO_2 \tag{4}$$

It was also observed that sodium and potassium were in carbonate form and not oxide at operating temperature of BLG.

Knowing whether the oxide or carbonate is stable, the thermodynamic stability of aluminates against sodium carbonate could be evaluated. Figure 6 is the result of this modeling in the form of ΔG of reaction versus temperature. Based on this figure, all candidate aluminates were stable against sodium carbonate and among them magnesium aluminate spinel was the most resistant one because it showed the highest ΔG of reaction with sodium carbonate.

More evaluation is necessary before recommending aluminates for the linings of high temperature gasifiers because, although they were resistant to sodium carbonate, they were

---

† M: Mg, Ba, Ca, Li, Na, K
$n_{(Mg, Ba, Ca)}=2$, $n_{(Li,Na,K)}=1$

corroded by sodium oxide. Thermodynamic modeling showed that sodium was stable in the form of sodium carbonate and sodium sulfide in the operating conditions of high temperature BLG, but field trials are necessary since the presence of water vapor in the gasifier or unstable operating conditions may lead to Na_2O formation in actual operation.

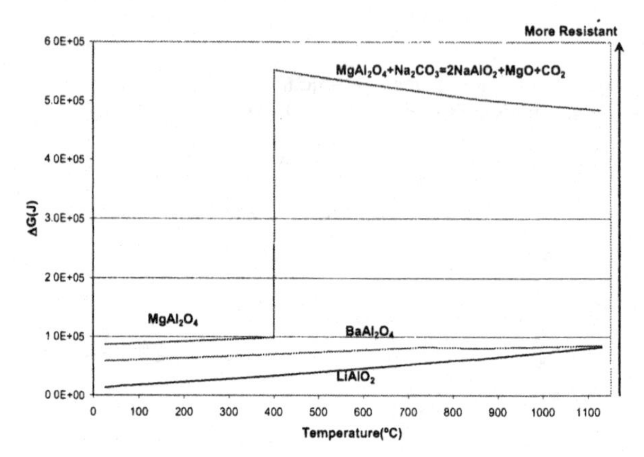

Fig. 6: ΔG of reactions between aluminates and sodium carbonate

Aluminates are not resistant to potassium containing compounds which were part of black liquor. The main reaction of the aluminates with potassium oxide was as follows:

$$^{\ddagger}M_nAl_2O_4 + K_2O \longrightarrow M_nO + 2KAlO_2 \tag{5}$$

The Gibbs free energy change of the reaction as a function of temperature is plotted in Figure 7 for lithium, barium and magnesium aluminate. It was observed that none of the aluminates resists potassium oxide. Among them, barium aluminate was the most resistant and magnesium aluminate was the least resistant. A reaction equation for aluminates with potassium carbonate was as follows:

$$M_nAl_2O_4 + K_2CO_3 \longrightarrow M_nCO_3 + 2KAlO_2 \tag{6}$$

The ΔG of the reactions between candidates and potassium carbonate as a function of temperature are plotted in Figure 8. Among the three aluminates, only lithium aluminate was resistant to potassium carbonate. Barium and magnesium aluminates were not resistant to potassium carbonate at the operating temperature of BLG.

It can be summarized that all three aluminates were resistant to sodium carbonate, but not sodium oxide. None of the aluminates were resistant to potassium oxide, but lithium aluminate was resistant to potassium carbonate.

---

‡‡ M: Mg, Ba, Li
$n_{(Mg, Ba)}=2$, $n_{(Li)}=1$

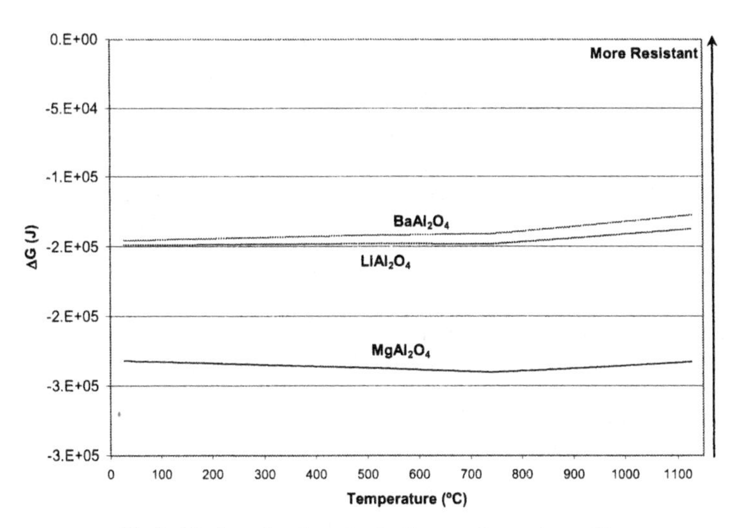

Fig. 7: ΔG of reactions between aluminates and potassium oxide

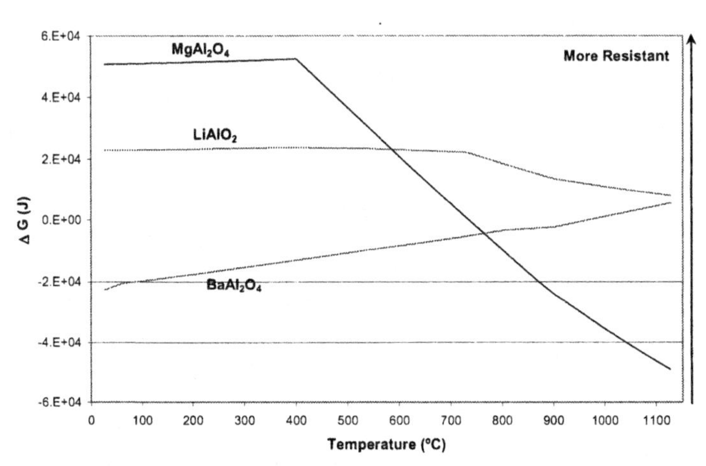

Fig. 8: ΔG of reactions between aluminates and potassium carbonate

Experimental analysis is necessary to verify the results of the thermodynamics. Sessile drop test will be employed to study the wetting behavior of the materials by black liquor as well as the resistance to corrosion reactions. This experimental work is in progress at University of Missouri-Rolla.

3. CONCLUSION

Worldwide growth of black liquor production as a new source of energy and electricity necessitates the development of new refractory materials resistant to harsh operating conditions of black liquor gasifiers. Current BLG systems use aluminosilicates refractories, which are not resistant to reaction with black liquor smelt. Widespread implementation of this technology requires the development of new refractory lining materials. Thermodynamic analysis showed that oxides such as magnesia, ceria and zirconia or aluminates such as barium and lithium aluminate may have satisfactory stability against black liquor smelt. Non-oxides such as SiC and Si_3N_4 were dissolved by black liquor smelt and were not candidates for this application.

4. REFERENCES

1. J. R. Keiser, R. A. Peascoe, and C. R. Hubbard, "Corrosion Issues in Black Liquor Gasifiers"; pp. 19 in Corrosion/2003 Conference Proceedings, NACE International, San Diego, CA, 2003.
2. L.L. Stigsson and B. Hesseborn, "Gasification of Black Liquor"; Section B, pp. 277-295 in International Chemical Recovery System Proceedings, Montreal Technical Section, CPPA, Toronto, Ontario, Canada, 1995.
3. C. L. Verrill, J. B. Kitto and J. A. Dickinson, "Development and Evaluation of a Low-Temperature Gasification Process for Chemical Recovery from Kraft Black Liquor"; pp. 1067-1078 in International Chemical Recovery Conference Proceedings, TAPPI Press, Tampa, FL, 1998.
4. E. Dahlquist, R. Jacobs, "Development of a Dry Black Liquor Gasification Process"; pp. 457-471 in International Chemical Recovery Conference Proceedings, TAPPI Press, Seattle, WA, 1992.
5. L. Stigsson, "Chemrec Black Liquor Gasification"; pp. 663-674 in International Chemical Recovery Conference Proceedings, TAPPI Press, Tampa, FL, 1998.
6. C. Brown, P. Smith, N. Holmblad, G. M. Christiansen, and B. Hesseborn, "Update of North America's First Commercial Black Liquor Gasification Plant"; pp. 33-49 in Engineering and Papermakers Conference Proceedings, TAPPI Press, Nashville, TN, 1997.
7. T. M. Grace, and W. M. Timmer, "A Comparison of Alternative Black Liquor Recovery Technologies"; Section B, pp. 269-B275 in International Recovery Conference Proceedings, TAPPI Press, Toronto, Ontario, Canada, 1995.
8. E. D. Larson and D. R. Raymond, "Commercializing Black Liquor and Biomass Gasifier/Gas Turbine Technology," *TAPPI J.*, **80** [2], 50-57 (1997).
9. S. Consonni, E. D. Larson, N. Berglin, "Black Liquor-Gasification/Gas Turbine Cogeneration"; pp. 1-9 in The American Society of Mechanical Engineers, Vol. 97-GT-273, ASME, 1997.
10. J. Gullichsen, H. Paulapuro, "Chemical Pulping", pp. B13-B18, Paper Making Science and Technology series book; Published in cooperation with the Finnish Paper Engineers' Association and TAPPI Book 6B, 1999.
11. R. A. Peascoe, J. R. Keiser, C. R. Hubbard and M. P. Brady, "Performance of Selected Materials in Molten Alkali Salts"; pp. 189-200 in 10th International Symposium on Corrosion in the Pulp and Paper Industry Proceedings, Technical Research Centre of Finland, Helsinki, Finland, 2001.

Mechanics

AN INVESTIGATION OF WETTABILITY, AND MICROSTRUCTURE IN ALUMINA JOINTS BRAZED WITH Ag-CuO-TiO_2

Jens T. Darsell
Washington State University
PO Box 642920
Pullman, WA 99164-2920

John S. Hardy, Jin Y. Kim, and K. Scott Weil
Pacific Northwest National Laboratory
P.O. Box 999
Richland, WA 99352

ABSTRACT

A silver-based joining technique referred to as reactive air brazing (RAB) has been recently developed for joining high temperature structural ceramic components of the type used in gas turbines, combustion engines, heat exchangers, and burners. It was found that additions of CuO to silver have a significant effect on the wettability and joint strength characteristics of the resulting braze on polycrystalline alumina substrates. More recently, it has been found that by adding as little as 0.5 mol % titania to these Ag-CuO brazes, the wettability of the RAB on alumina surfaces is further enhanced. The results of wettabilty measurements of Ag-CuO-TiO_2 RAB compositions on alumina will be presented along with the microstructural characterization of Ag-CuO-TiO_2 braze joints in alumina.

INTRODUCTION

Ceramic materials have found many uses for high temperature applications because of their oxidation resistance, compressive strength, and wear resistance at high temperatures. These applications include turbine blades, chemical sensors as well as high temperature ceramic based solid oxide fuel cells. Unfortunately, ceramic materials can be quite difficult and expensive to machine due to their high hardness values, low thermal conductivities, and brittle nature. Therefore, ceramic components are often slurry cast and sintered or hot pressed as close to the final shape as possible. However, these processing techniques are not always feasible with large parts or complicated geometries. An alternative is to join simple-shaped pieces to build to the final, more complex part.

Several options that are available for joining ceramics include glass bonding and active metal brazing. Glass bonding is typically limited by the softening point of the glass, which dictates the maximum operating temperature of the joint. Active metal brazing employs a braze alloy containing one or more reactive elements, such as Ti, to modify the ceramic surface in order to make wetting more favorable. Active metal brazing is typically conducted under inert atmospheres or in vacuum in order to control the extent of the reaction and to avoid oxidizing the reactive species. It has been shown that complete oxidation of an active metal braze deteriorates the performance of the braze[1]. A new technique that has shown promise in ceramic joining is Reactive Air Brazing (RAB). In RAB, an inert filler material is used, such as Ag, Au, or Pt, along with an appropriate oxidizable alloying agent such as Cu. Brazing is conducted directly in air without need of a cover gas or surface cleaning fluxes. In this way, the alloying agent oxidizes during the heating process and forms a reactive product that modifies the ceramic surface and improves the wetting characteristics of the braze.

One particular system of interest in the development of RAB has been the binary Ag-CuO. Meier et al. have shown that increasing the CuO content significantly improves the wetting behavior of Ag-CuO on alumina[2] in an inert environment. The Ag-CuO system was demonstrated to be capable of joining alumina[3, 4, 5] and to have improvements in wetting behavior with CuO additions[5] in air. In addition, investigation of the copper oxide - titanium dioxide phase diagram indicates that a eutectic reaction occurs at 919 °C for X_{TiO2} = 16.7 mol %[6]. Based on this, Weil and Hardy explored the effects of adding TiO_2 to the Ag-CuO braze system[7]. Their study focused on the wetting of $(La_{0.6}Sr_{0.4})(Co_{0.2}Fe_{0.8})O_3$ and Fecralloy (22% Cr, 4.8% Al, 0.3% Si, 0.3% Y, bal. Fe) substrates by Ag-CuO-TiO_2 as a potential means of brazing solid state oxygen separation devices. They determined that the addition of TiO_2 to the Ag-CuO braze system can significantly reduce the contact angle on both substrates. The goal of our present study is to understand the effect of TiO_2 in the Ag-CuO braze system on the wettability of a model substrate, alumina.

EXPERIMENTAL

Materials

Sessile drop experiments were performed on polycrystalline alumina substrates (Al-23, Alpha Aesar, Ward Hill, MA 01835), 98% dense and 99.7 % pure containing a small amount of silicate material. These disks are 50 mm in diameter and 6 mm thick. The surface of the disk to be brazed was polished to a 1 μm finish using water based diamond suspensions. The disks were then cleaned with acetone followed by rinsing with propanol and ethanol. The disks were then air dried and heated in a static air furnace to 600 °C and held for four hours to burn off any organic contamination.

Pellets were fabricated by dry mixing powders of silver (spherical, 0.5-1 μm average particle size, 99.9% pure by metals basis, Alpha Aesar, Ward Hill, MA 01835), copper (1-1.5 micron average particle size, 99% pure by metals basis, Alpha Aesar, Ward Hill, MA 01835) and titanium hydride (-325 mesh, 98% pure, Aldrich, Milwaukee, WI). For sessile drop experiments 0.84 g of powders were pressed into approximately 7 mm diameter by 10 mm tall pellets. Titanium hydride decomposes at approximately 450 °C to form titanium which oxidizes to form TiO_2 during the experiment.

Characterization

Sessile drop experiments were performed in a static air muffle furnace modified with a quartz window that allowed viewing of the sample while heating. The braze pellets were placed on the substrate in the furnace. Samples were first heated to a temperature of 900 °C at 30 °C/min and held for 15 min. This was followed by a heating rate of 10 °C/min, with holds for 15 minutes at 950, 1000, 1050, and 1100 °C. A video camera with a zoom lens was utilized to record the profile of the specimens during the heating cycle. Ulead™ software was used to convert portions of the video tape to digital images which were used to measure the contact angle between the braze pellet and the alumina substrate. Microscopic analysis was performed on polished cross sections of the wetting samples using a scanning electron microscope (SEM, JEOL, JSM-5900) equipped with an energy dispersive X-ray (EDX) detector and analysis system. Samples were mounted in epoxy and polished with diamond suspension to a final grit size of 0.25 μm. Samples were coated with carbon to avoid charging.

Table I. Braze compositions employed in this study

Braze I.D.	Ag Content (mol %)	CuO Content (mol %)	TiO_2 Content (mol %)
Ag0.5Ti	99.5	0	0.5
Ag1Ti	99	0	1
99Ag1Cu	99	1	0
99Ag1Cu0.5Ti	98.5	1	0.5
99Ag1Cu1Ti	98	1	1
99Ag1Cu2Ti	97	1	2
98.6Ag1.4Cu	98.6	1.4	0
98Ag2Cu	98	2	0
98Ag2Cu0.5Ti	97.5	2	0.5
98Ag2Cu1Ti	97	2	1
96Ag4Cu	96	4	0
96Ag4Cu0.5Ti	95.5	4	0.5
96Ag4Cu1Ti	95	4	1
96Ag4Cu2Ti	94	4	2
92Ag8Cu	92	8	0
92Ag8Cu0.5Ti	91.5	8	0.5
92Ag8Cu1Ti	91	8	1
90Ag10Cu	90	10	0
84Ag16Cu0.5Ti	83.5	16	0.5
84Ag16Cu1Ti	83	16	1
84Ag16Cu2Ti	82	16	2
80Ag20Cu	80	20	0
70Ag30Cu	70	30	0
66Ag34Cu0.5Ti	65.5	34	0.5
66Ag34Cu2Ti	64	34	2
60Ag40Cu	60	40	0
50Ag50Cu	50	50	0
40Ag60Cu	40	60	0
40Ag60Cu	40	60	0
31Ag69Cu	30.65	69.35	0
31Ag69Cu0.5Ti	30.15	69.35	0.5
31Ag69Cu2Ti	28.65	69.35	2
20Ag80Cu	20	80	0

RESULTS AND DISCUSSION

Wetting

Shown in Figure 1 is a comparison of the wetting behavior for 96 mol %-Ag 4 mol % CuO with additions of 0, 0.5, 1, and 2 mol % TiO_2. The data indicate that the contact angle of Ag-CuO decreases significantly with the addition of TiO_2. However, the difference in contact angle for brazes with 0.5, 1, and 2 mol % TiO_2 is within the experimental error of ±3°. Apparently 0.5 mol % TiO_2 is sufficient to improve wetting and further additions do not increase the effect. Figure 2 displays the contact angle at 1100 °C as a function of copper oxide content for both the binary and ternary compositions. It can be seen that the addition of TiO_2 lowers the contact angle from the baseline of Ag-CuO in the range of approximately 66-99 mol % Ag. The contact angle decreased by approximately 25° for the 99Ag1Cu0.5Ti sample over the 99Ag1Cu sample. However, this difference in wetting behavior is reduced to within experimental error for braze compositions containing less than 66 mol % Ag; i.e. with respect to wetting, at CuO contents of $\geq$ 34 mol %, the addition of TiO_2 offers no benefit over the corresponding binary braze. Alternatively, at low CuO-content, a detrimental effect is observed. For example, the 99Ag1Cu0.5Ti sample displays a relatively low contact angle of 33°, while the 99Ag1Cu1Ti sample exhibits a much higher contact angel of 66°, and the 99Ag1Cu2Ti composition displays a still higher contact angle of 70°, which is almost identical to that of the analogous binary, 99Ag1Cu. This data suggests that in addition to CuO-content, the ratio of TiO_2 to CuO is an important variable defining the wetting behavior of the ternary braze.

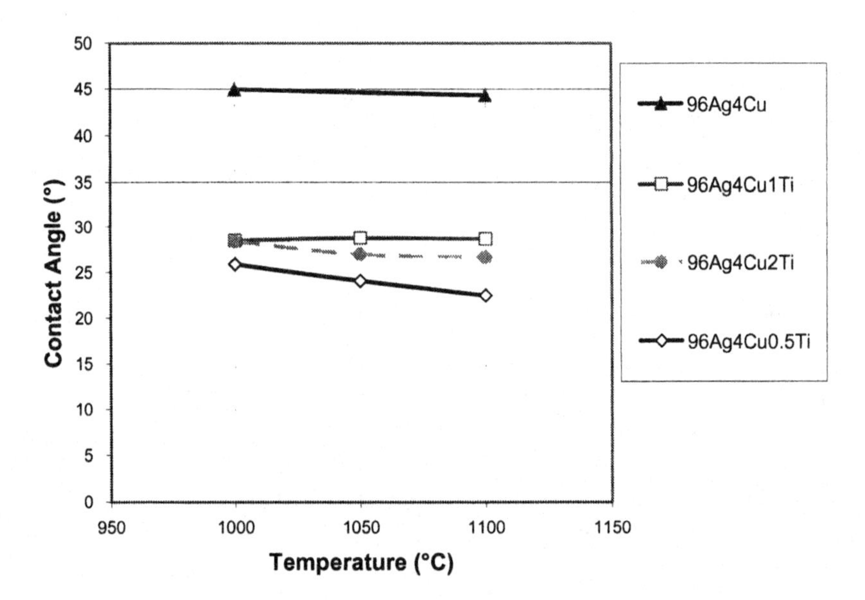

Figure 1 Contact angle as a function of temperature for samples with 96 mol % Ag-4 mol % CuO with additions of 0.5, 1.0, and 2.0 mol % TiO_2. Samples were held at given temperatures for 15 minutes.

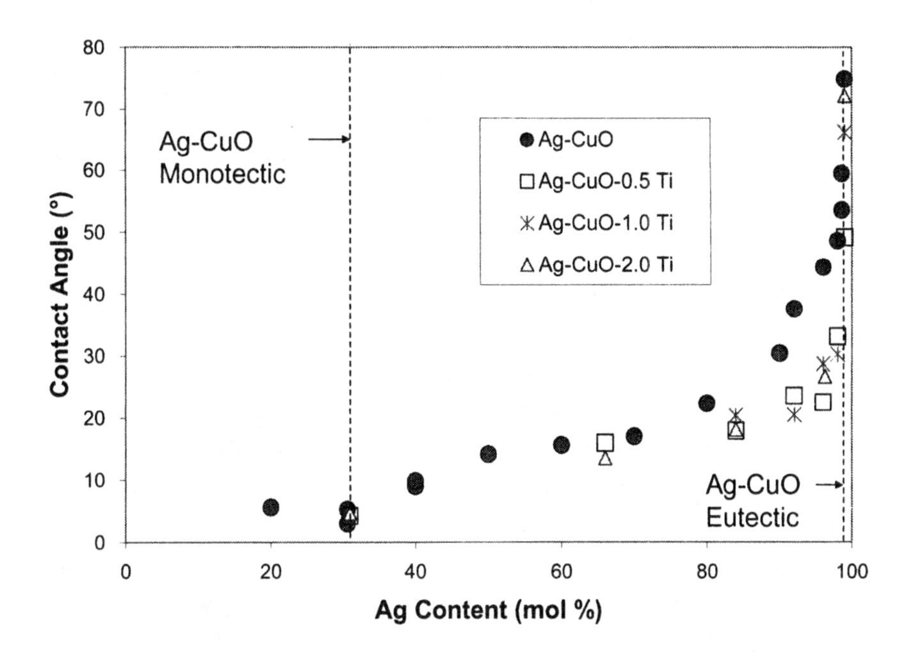

Figure 2 Contact angle as a function of Ag content for Ag-CuO with additions of 0.5, 1.0, and 2.0 mol % TiO_2 after a final hold temperature of 1100 °C for 15 minutes.

Microstructure

Shown in Figures 3(a) – (d) are cross-sectional SEM images of the braze/alumina interfaces in sessile drop specimens produced with 96Ag4Cu, 96Ag4Cu2Ti, 70Ag30Cu, and 66Ag34Cu2Ti. As seen in Figure 3(a), away from this interface the solidified 96Ag4Cu sessile drop displays a silver rich region containing copper oxide precipitates measuring on average ~1 – 3 μm in size. At the interface with alumina, the copper oxide forms a 10 μm thick semi-continuous layer that appears to be in direct contact with the substrate. Quantitative EDX analysis indicates that the precipitates and interfacial oxide layer contain oxygen and copper. Traces of copper are also found within the alumina substrate along grain boundaries, indicating possible diffusional migration of copper into the alumina. The effect of 2 mol % TiO_2 addition on this Ag-CuO composition is shown in Figure 3(b). Some grain boundary dissolution is apparent by the infiltration of the braze filler into the sub-surface of the alumina. EDX analysis indicates that this intergranular phase is composed of Al, Cu, and O. Interestingly, titanium was not found in this region or at the original braze/substrate interface, but rather within precipitates in the bulk and at the top surface of the solidified drop, as shown in Figure 4.

In the 70Ag30Cu/alumina sample, a fully continuous layer of copper oxide forms in contact with the alumina substrate and measures on average approximately 50 μm thick. As shown in the high magnification image in Figure 3(c), a substantial amount of silver is found within this oxide band. Based on the phase diagram for Ag-CuO[8,9], at the composition for this braze two immiscible liquids are expected to form at 1100 °C, one rich in CuO and the other rich in silver. We propose that when segregation takes place, the oxide-rich liquid will preferentially wet the substrate because of its higher CuO content and, therefore, lower expected interfacial energy with Al_2O_3. As the specimen is further cooled, copper oxide will first precipitate out on the surface, then below the eutectic point, both silver and additional copper oxide will precipitate onto this proeutectic layer. This possibly explains not only the prevalence of CuO at the droplet/substrate interface, but also the occurrence of silver within. The effect of adding TiO_2 to a sample containing a large amount of copper oxide can be seen in Figure 3(d). A silver rich region with light gray particles containing Cu and O can be found in the Ag matrix as well as at the interface. Darker gray particles containing Al, Cu, and O, as well as a mixed phase containing Cu, Ti, Al, and O are also found at the interface.

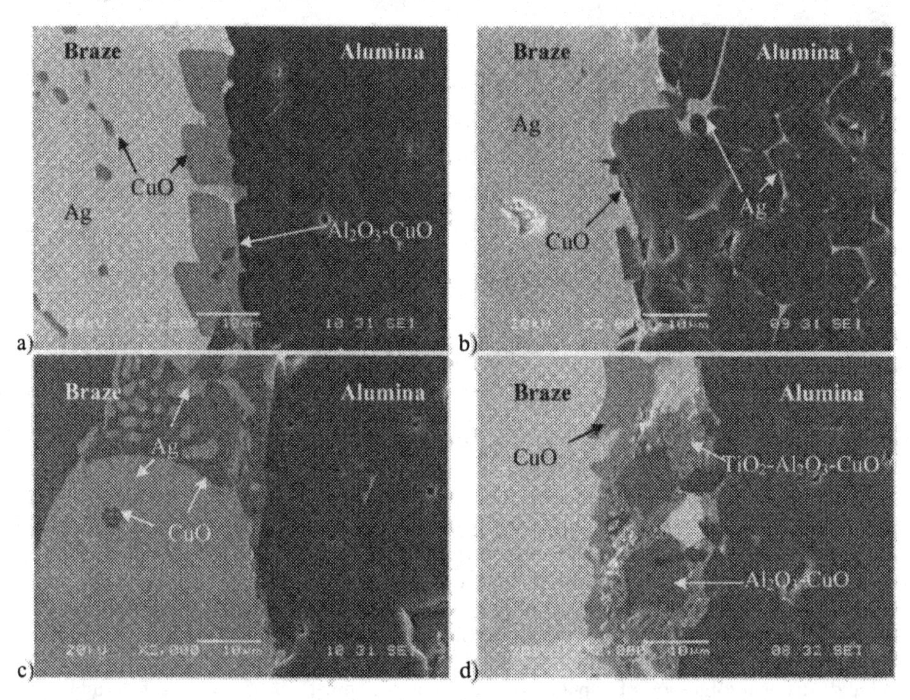

Figure 3 Scanning electron micrograph at 2000X, comparison of the braze/alumina interface formed with a) 96Ag4Cu, b) 96Ag4Cu2Ti, c) 70Ag30Cu, and d) 66Ag34Cu2Ti. All samples were held at 1100 °C for 15 minutes.

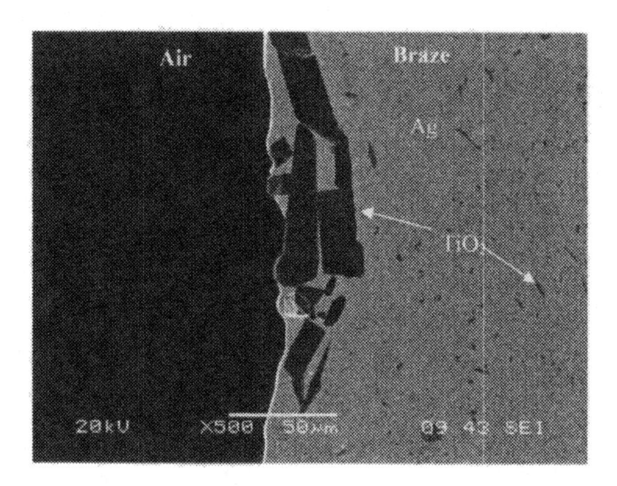

Figure 4 Scanning electron micrograph at 500X, of the 96Ag4Cu2Ti/alumina sessile drop sample shows precipitates containing Ti at the top of the braze surface of solidified drop (air/braze interface).

CONCLUSIONS

The addition of TiO_2 to the Ag-CuO braze system was studied by observing the wetting behavior and microstructure of the resultant braze on alumina. It was found that the addition of amounts as small as 0.5 mol % of TiO_2 could significantly decrease the contact angle for the Ag-CuO system on alumina in the range of 66-99 mol % Ag. This effect is most apparent in the brazes with large amounts of Ag and not apparent bellow 34 mol % Ag. No benefit was seen by adding 1 or 2 mol % TiO_2 over 0.5 mol % on the wetting behavior, although the ratio of CuO to TiO_2 may be an important factor. Microscopy shows that all of the samples wet the alumina substrate well. The addition of TiO_2 causes increased grain boundary dissolution of the alumina. Particles containing Ti were found in the bulk of the braze and at the braze/air interface for sample containing small amounts of CuO. Samples containing large amounts of CuO formed a mixed Cu, Ti, Al, and O phase at the braze/alumina interface in addition to CuO precipitates and a Cu, Al, and O phase.

ACKNOWLEDGEMENTS

The authors would like to thank Nat Saenz, Shelly Carlson, and Jim Coleman for their assistance in polishing a portion of the wetting samples and conducting the metallographic and SEM analysis work. This work was supported by the U.S. Department of Energy, Office of Fossil Energy, Advanced Research and Technology Development Program. J. Darsell would like to thank his thesis advisors at the School of Mechanical and Materials Engineering at Washington State University, Pullman WA, Amit Bandyapodyay and Susmita Bose. The Pacific Northwest National Laboratory is operated by Battelle Memorial Institute for the United States Department of Energy (U.S. DOE) under Contract DE-AC06-76RLO 1830.

REFERENCES

[1] J.P. Rice, D.M. Paxton, and K.S. Weil, "Oxidation Behavior of a Commercial Gold-Based Braze Alloy for Ceramic-to-Metal Joining," *Ceramic Engineering and Science Proceedings*, **23** [3] 809-816 (2002).

[2] A.M. Meier, P. R. Chidambaram, G.R. Edwards, "A Comparison of the Wetability of Copper-Copper Oxide and Silver-Copper Oxide on Polycrystalline Alumina," *Journal of Materials Science*, **30** [19] 4781-86 (1995).

[3] C.C. Shüler, A. Stuck, N. Beck, H. Keser, and U. Täck, "Direct Silver Bonding – An Alternative for Substrates in Power Semiconductor Packaging," *Journal of Materials Science: Materials in Electronics*, **11** [3] 389-96 (2000).

[4] K.M. Erskine, MS Thesis, Alfred University, Alfred, NY (1999).

[5] J.Y, Kim, and K.S. Weil, "Development of a Copper Oxide-Silver Braze for Ceramic Joining," *Ceramic Transactions: Advances in Joining of Ceramics*, Vol. 138, pp.119-132, Edited by C.A. Lewinsohn, M. Singh, and R. Loehman. The American Ceramic Society, (2003).

[6] F-H. Lu, F-X. Fang, and Y-S. Chen, "Eutectic Reaction Between Copper and Titanium Dioxide," *Journal of the European Ceramic Society*, **21** [8] 1093-1099 (2001)

[7] K.S. Weil and J.S. Hardy, "Brazing a Mixed Ionic/Electronic Conductor to an Oxidation Resistant Metal," *Advances in Joining of Ceramics, Ceramic Transactions*, Volume 138

[8] Z.B. Shao, K.R. Liu, and L.Q. Liu, "Equilibrium Phase Diagrams in the Systems PbO-Ag and CuO-Ag," *Journal of the American Ceramic Society*, **76** [10] 2663-64 (1993)

[9] H. Nishiura, R.O. Suzuki, K. Ono, and L.J. Gauckler, "Experimental Phase Diagram in the Ag-Cu_2O-CuO System," *Journal of the American Ceramic Society*, **81** [8] 2181-87 (1998)

AN ENGINEERING TEST USEFUL IN DEVELOPING GLASS SEALS FOR PLANAR SOLID OXIDE FUEL CELLS

K.S. Weil, J.E. Deibler, J.S. Hardy, D.S. Kim, G-G. Xia, and C.A. Coyle
Pacific Northwest National Laboratory
P. O. Box 999
Richland, WA 99352

ABSTRACT

Developing reliable, hermetic ceramic-to-metal seals has become a key issue to the future success of planar solid oxide fuel cell (pSOFC) technology. In the course of designing and fabricating operating pSOFC stacks at Pacific Northwest National Laboratory over the past several years, we have relied on a simple, but extremely useful seal testing technique that allows us to screen through the numerous variables involved in developing glass seals. Here we discuss the design of the test apparatus and the procedure we employed when using the test to analyze the effects of various processing parameters on the hermeticity and durability of our glass seals.

INTRODUCTION

Solid oxide fuel cells function because of an oxygen ion gradient that develops across the yttria stabilized zirconia (YSZ) electrolyte membrane via ionic transport when one side is exposed to an oxygen-rich environment, such as air, and the other to reducing gas. In order to maintain this gradient, and thereby maximize the performance of the device, the electrolyte and the joint that seals this membrane to the device chassis must be hermetic. That is the YSZ layer must be dense, must not contain interconnected porosity, and must be connected to the rest of the device structure with a high temperature, gas-tight seal. With planar SOFCs, recent advances in the tape casting of thin, anode-supported ceramic bilayers have successfully addressed the first two issues [1]. The remaining challenge is to develop a consistent method of joining the ~10μm thick electrochemically active YSZ electrolyte to the metallic structural component such that the resulting seal is hermetic, rugged, and stable under both thermal cycling and continuous long-term high-temperature operation.

There are essentially two standard methods of sealing planar SOFCs, either by forming a rigid joint or by constructing a compressive "sliding" seal. Our work has focused on the former technique - specifically, variations on the traditional glass sealing approach. While glass joining is a cost-effective and relatively simple method of bonding ceramic and metal parts, there are several limitations that must be addressed, including: (1) the softening point of the glass limits the maximum operating temperature to which the joint may be exposed and (2) the final seal is brittle, non-yielding, and particularly susceptible to fracture when exposed to tensile stresses. Because of these concerns, it is imperative in glass-sealed stack designs that the coefficient of thermal expansion (CTE) of each joined component, i.e. the ceramic cell, the seal, and the metal separator, be approximately equal to minimize the build up of residual stresses within the joint. Currently only a handful of high temperature glass compositions satisfy this requirement [2]. Additionally as outlined in Table 1, there are a number of other variables that can affect the performance of this type of seal as well.

The relationship between these processing/material variables and seal performance (i.e. hermeticity and strength) can be investigated using traditional methods of mechanical analysis such as tensile and torsion testing. If conducted carefully, these tests provide a highly accurate, quantitative measure of joint strength. However in practice these techniques are not especially conducive to a development program, where rapid specimen fabrication, testing, seal modification, re-testing, etc. are essential to success. Specifically, test specimen preparation and alignment of the specimen in the test fixture can be particularly difficult and laborious. An alternative method of testing was sought that would offer a much faster turn-around time and facilitate the examination of a large number of materials, processing, and performance parameters. Here, we describe a modified version of a standard rupture testing technique that was used as the primary qualification tool in our screening effort. The test is conducted by placing a sealed disk specimen in the test fixture and pressurizing the backside of the sample until seal rupture occurs. That is, the specimen is subjected to an accelerated stress test, using air pressure to generate high levels of stress within the seal. The details of the test apparatus and procedure are outlined below along with example results that have influenced our stack design.

Table 1 Materials, Processing, and Performance Variables Involved in pSOFC Glass Seal Development

Material Variables	Processing Variables	Performance Variables
Glass composition	Binder burn-out conditions	Thermal exposure
Ratio of BaO, Al_2O_3, and SiO_2	Burn-out temperature	Temperature
Minor alloying agents: L_2O_3, CeO_2, and CaO	Time at temperature	Time at temperature
Second Phase Additions		Ambient gas atmosphere
Metal Substrate Effects	Joining temperature	Thermal shock
Scale composition: Cr_2O_3 vs. Al_2O_3	Temperature uniformity during sealing	Magnitude of ΔT/cm across
CTE effects	Glass viscosity as a function of soak temperature	the seal due to rapid
Use of minor alloing agents: Mn , Ti, etc.		heating/cooling
Ceramic Substrate Effects	Time at joining temperature	Thermal cycling
YSZ composition: 3, 5, and 8% Y_2O_3 additions	Glass viscosity as a function of soak time	Rate of heating/cooling
CTE effects: anode composition		Number of thermal cycles
Use of substrate surface modifiers	Uniformity and magnitude of joining pressure	
Coatings	Glass viscosity as a function of sealing pressure	
Surface geometry		
	Rate of heating during joining	

EXPERIMENTAL

Materials

As shown in Figure 1, the test employs essentially a miniaturized version of the main fuel cell component as the test specimen. A 25mm diameter ceramic bilayer is glass sealed directly to a metal washer, of the same composition as the frame used in the pSOFC stack, that measures 44mm in outside diameter with a 15mm diameter concentric hole. Like the actual ceramic pSOFC cell, the anode-supported bi-layer coupons consist of NiO-5YSZ as the anode and 5YSZ as the electrolyte in their prototypic thicknesses. The bilayer coupons were fabricated by tape casting and co-sintering techniques developed at Pacific Northwest National Laboratory [3]. After heat treatment, the finished bi-layer components measured nominally 25mm in diameter by 600μm in thickness, with an average electrolyte thickness of ~8μm.

Eight different ferritic stainless steel interconnect candidates were chosen for initial testing, five of which form a protective chromia scale upon high temperature air exposure and three that form an alumina scale, as listed in Table 2. With the exception of Durafoil, each steel was a procured as 300μm thick sheet in the as-annealed condition. The Durafoil was obtained as 50μm thick sheet. The flat washer-shaped specimens were cut from the sheets via electrical discharge machining, ultrasonically cleaned in acetone for 10 minutes, and wiped with methanol prior to use.

The glass seal composition employed in this study, designated as G-18, was an-house designed barium calcium aluminosilicate based glass originally melted from the following mixture of oxides (by weight percent): 56.4% BaO, 22.1% SiO_2, 5.4% Al_2O_3, 8.8% CaO, and 7.3% B_2O_3 [4]. The G-18 powder was milled to an average particle size of ~20μm and mixed with a proprietary binder system to form a paste that could be dispensed onto the substrate surfaces at a uniform rate of 0.075g/linear cm using an automated syringe dispenser. In this manner, the glass paste was dispensed onto the YSZ side of the bilayer discs. Each disc was then concentrically positioned on a washer specimen, loaded with a 50g weight, and heated in air under the following sealing schedule: heat from room temperature to 850°C at 10°C/min, hold at 850°C for one hour, cool to 750°C at 5°C/min, hold at 750°C for four hours, and cool to room temperature at 5°C/min.

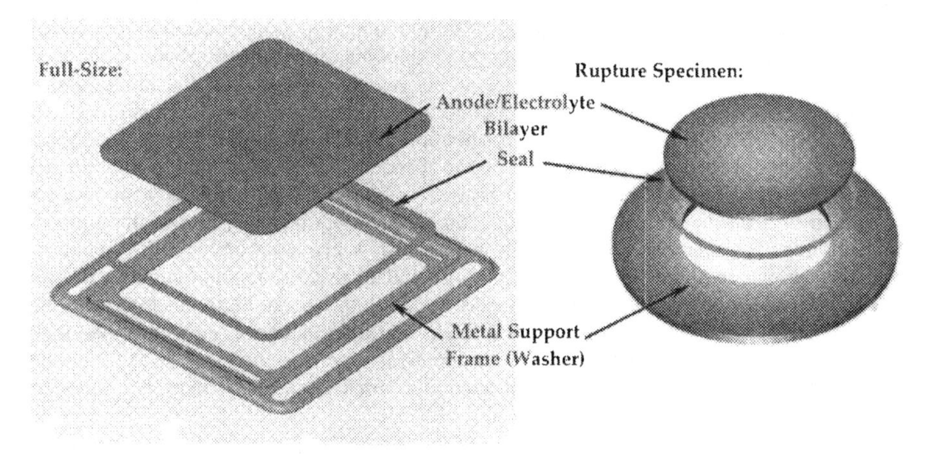

Figure 1 A comparison of the the full-size SOFC window frame component to the rupture test specimen.

Table 2 Ferritic Stainless Steel Composition

	Nominal composition, wt%											
Alloy	Ni	Cr	Fe	Mo	C	Mn	Si	Al	P	S	N	Others
Al 29-4	0.15	29.0	Balance	4.0	0.01	0.3	0.2	-	0.025	0.02	0.015	
Alpha-4 (Durafoil)	-	21.0	Balance	-		-	-	6.3	0.02	0.02		0.1 Ce + La
Crofer-22 APU	-	22.8	Balance	-	0.005	0.45	-	-	0.016	0.002	0.01	0.08 Ti + 0.06 La
E-Brite 26-1	0.09	26.0	Balance	1.0	0.001	0.01	0.025	-	0.02	0.02	0.01	0.03 Cu
Fecralloy	-	22.0	Balance	-	0.03	-	-	5.3	0.02	0.02	0.01	0.10 Y + 0.01 Zr
430	-	17.0	Balance	-	0.12	1.0	1.0	-	0.04	0.03	0.25	
Aluminized 430	-	17.0	Balance	-	0.12	1.0	1.0	5.0	0.04	0.03	0.25	
446	-	25.0	Balance	-	0.020	1.50	1.00	-	0.04	0.03	0.25	

Testing

One can envision several different mechanisms by which a rigid glass pSOFC seal could fail during operation, including: (1) fracture initiated by a difference in the incoming fuel and air stream pressures across the membrane and seal, (2) failure due to out-of-plane bending stresses or high tensile stresses generated by rapid thermal cycling, and (3) failure due to a loss of CTE matching as the glass devitrifies during the first few hours of exposure at operating temperature [5]. Given these modes of failure, our screening methodology was to test various pSOFC joining systems, i.e. combinations of ceramic/metal substrates and sealing materials, under sequentially more aggressive conditions in an effort to identify the most promising sealing candidates. In this way, seals were first rupture tested in the as-joined condition. Joining systems that were considered viable were then re-tested after prolonged thermal exposure and/or upon multiple heating and cooling cycles. Below we discuss the results from our initial as-joined tests.

A schematic of the experimental set-up used in rupture testing is illustrated in Figure 2(a) and a photo of the device is shown in Figure 2(b). The test sample is placed within a fixture that consists of a bottom and top flange, a coupling that secures and centers the two flanges, and an o-ring that is squeezed against the bottom surface of the washer. We found that the dimensions of the two flanges and the o-ring relative to those of the specimen and the amount of torsion used to compress the o-ring can play a role in defining the rupture stress of the sealing specimen. Thus, these parameters were held fixed at the values listed in Table 3 so that batch-to-batch comparisons could be made. Compressed air is used to pressurize the back-side of the washer specimen up to a maximum rated pressure of 150psi. A digital regulator allows the pressure behind the joined bilayer disk to be slowly increased to a given set point. This volume of compressed gas can be isolated between the specimen and a valve, making it possible to identify a leak in the seal by a decay in pressure. In this way, the device can be used to measure the hermeticity of a given seal configuration without causing destructive failure of the seal. Alternatively, by increasing the pressure to the point of specimen rupture, we can measure maximum pressure that the specimen can withstand. A minimum of six specimens was tested for each joining condition.

(a)

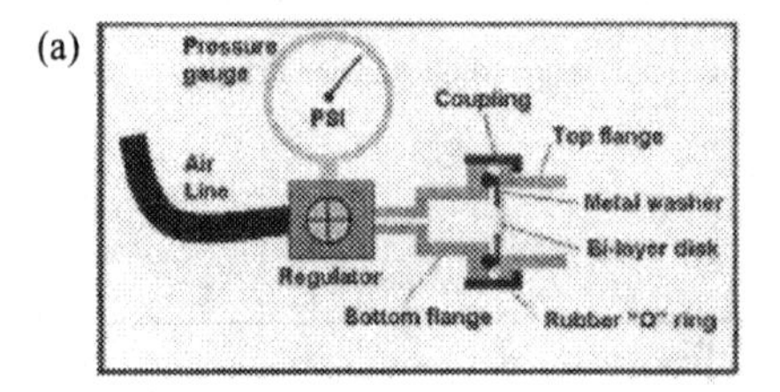

(b)

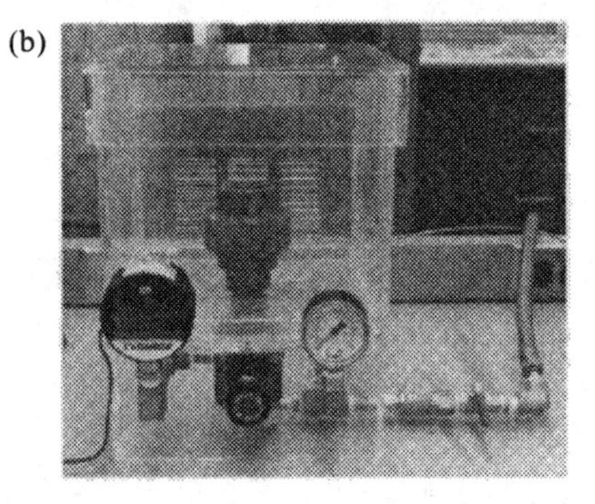

Figure 2 (a) A schematic of the rupture test apparatus. (b) A photo of the test aparatus.

It is important to recognize that the differential pressure that arises across each individual PEN (and therefore each PEN to frame seal) in a functional pSOFC stack is quite small under typical steady-state operating conditions. That is, the rupture strength test places the PEN, seal, and metal interconnect material under a highly exaggerated stress condition relative to prototypic operation. By doing this, the test allows us to identify the weak link in the sealing system, i.e. the ceramic substrate, the sealing glass, the metal substrate and associated oxide scale, or any of the interfaces in between, so that the seal can potentially be improved in the next round of development. Ideally for the purposes of quantitative comparison one would like the stress in the seal to be either pure shear or pure tension at failure. Unfortunately, the stress state in the rupture test specimen is mixed-mode. However as will be discussed, we found that the range of rupture pressures at failure for a given batch of specimens fabricated and tested under identical conditions was in general fairly small and that the average rupture pressure for the batch provides a useful figure of merit that could be used to compare with other batches joined or tested under a different set of conditions.

Table 3 Specimen and Fixture Dimensions

Component	Dimensions		
Anode/electrolyte bilayer disk	25mm O.D.		600μm thick
Metal substrate	44mm O.D.	15mm I.D.	300μm thick
Viton O-ring	35mm O.D.	29mm I.D.	
Top Flange	46mm O.D.	29mm I.D.	
Bottom Flange	46mm O.D.	24mm I.D.	
Compression		15 in-lbs torsion	

After rupture testing, we carried out further analysis to determine why failure takes place and how joining and sealing could potentially be improved. Microstructural analysis of the joints after rupture testing was conducted on polished cross-sectioned samples using a JEOL JSM-5900LV scanning electron microscope (SEM) equipped with an Oxford energy dispersive X-ray analysis (EDX) system that employs a windowless detector. In order to avoid electrical charging on the samples, they were carbon coated and grounded. Elemental profiles were determined across the joint interfaces in the line-scan mode.

RESULTS AND DISCUSSION

Results from our initial study that measured rupture strength as a function of metal interconnect composition are plotted in Figure 3. Rupture strength is measurably higher when an

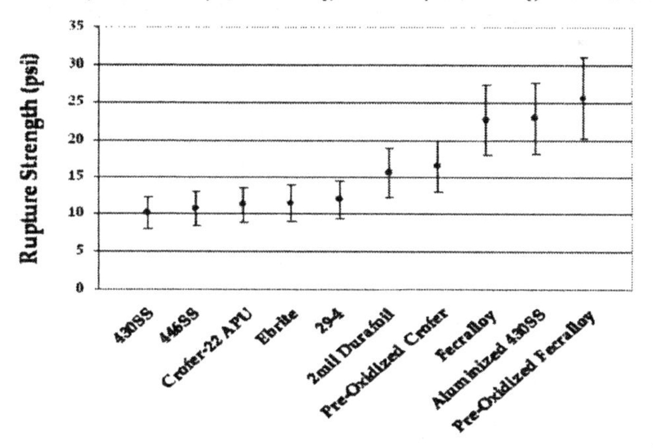

Figure 3 The rupture strength of as-joined PEM/glass/interconnect specimens as a function of metal substrate composition.

alumina-forming alloy is employed in the seal instead of a chromia-former. Microstructural analysis suggests that the root cause for this difference lies at the interface between the glass and the metal substrate. Shown in Figures 4(a) and (b) are representative cross-sectional micrographs of the glass/metal interface for a chromia-former, 446 stainless steel, and an alumina-former, FeCrAlY. Corresponding EDX analysis of each of the regions labeled in the two figures are provided in Table 4 (measured at 25KeV employing standards in the analysis).

Five microstructurally distinct zones are readily observed in the micrograph of the G-18/446 joint shown in Figure 4(a), moving from point 5 to point 1: (a) the bulk 446 stainless steel, (b) a chromia scale that forms on the surface of the alloy during joining, (c) a ~10μm thick reaction zone between the chromia scale and the glass, (d) an adjacent glass-ceramic region that is measurably depleted in barium, and (e) the devitrified glass-ceramic in the bulk portion of the seal. EDX analysis conducted on Point③ indicates that the primary phase in the interfacial reaction zone is $BaCrO_4$. Barium chromate has an orthorhombic crystal structure and displays the following coefficients of thermal expansion over a temperature range of 20 – 813°C: $\alpha_a = 16.5 \times 10^{-6}\ K^{-1}$, $\alpha_b = 33.8 \times 10^{-6}\ K^{-1}$, and $\alpha_c = 20.4 \times 10^{-6}\ K^{-1}$ [7], which differ substantially from the corresponding average CTEs for the devitrified glass and the 446, $11.8 \times 10^{-6}\ K^{-1}$ and $12.6 \times 10^{-6}\ K^{-1}$, respectively. We suspect that the mismatch is high enough to generate a large residual stress along this interface upon cooling, and thereby weaken the joint measurably during subsequent rupture testing.

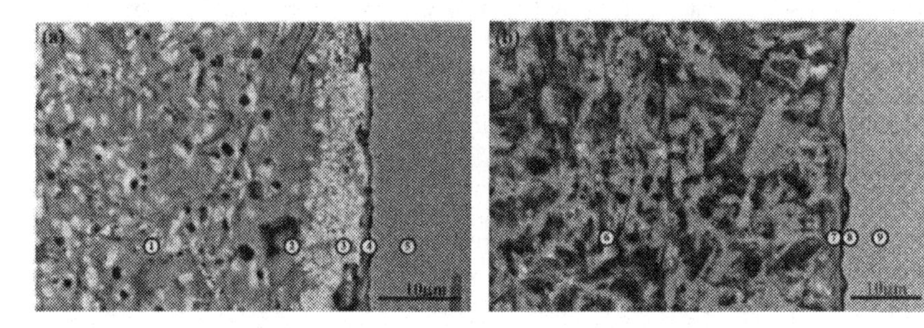

Figure 4 (a) The glass/metal interface against a representative chromia former, 446. (b) The glass/metal interface against a representative alumina former, fecralloy.

Table 4 Local EDS Results for Points Marked in Figures 4(a) and (b)

	Composition, at%						
Element	O	Al	Si	Ca	Cr	Fe	Ba
Point 1	57.4	3.8	18.2	2.6	-	-	18.0
Point 2	55.9	19.2	19.6	1.9	-	-	3.4
Point 3	65.1	1.9	4.2	0.6	13.3	0.2	14.7
Point 4	57.8	0.8	1.9	1.0	35.9	0.6	2.0
Point 5	-	-	-	-	23.3	76.3	-
Point 6	55.9	5.4	19.6	1.3	-	-	17.8
Point 7	59.0	14.8	16.3	1.3	1.7	0.3	6.6
Point 8	52.9	36.8	0.1	-	6.7	3.5	-
Point 9	-	4.3	0.5	-	21.6	73.6	-

In contrast, the FeCrAlY exhibits an oxidation rate that is nearly two orders of magnitude lower than 446 [6]. Thus during joining, FeCrAlY forms a much thinner oxide scale, composed primarily of alumina, and subsequently exhibits a much thinner reaction zone between the bulk glass and the scale. From the EDX analysis, the composition of the interfacial reaction zone corresponds to celsian, or $BaAlSiO_4$. This compound commonly assumes one of two crystal structures during formation, either a hexagonal crystal structure, referred to as hexacelsian with an average CTE of 8.2 x 10^{-6} K^{-1}, and a monoclinic structure, called monocelsian with an average CTE of 2.7 x 10^{-6} K^{-1}. Although monocelsian is thermodynamically the more stable phase, hexacelsian typically forms preferentially due to more favorable reaction kinetics [7]. We assume this holds true here as well, at least in the as-joined condition. Thus in cross-section, the joining sample in Figure 4(b) consists of four microstructurally distinct regions, moving from points 9 to 6: (a) the bulk alloy, (b) a very thin alumina scale, (c) a ~1µm thick celsian reaction zone, and (d) the bulk devitrified glass. Because the CTE of the hexacelsian is relatively close to that of the stainless steel and the reaction zone is quite thin, thermal expansion mismatch is not as problematic as in the case of $BaCrO_4$ formation and the resulting joint is stronger.

Post-test failure analysis of the rupture specimens confirms the source of fracture. Shown in Figures 5(a) and (b) are examples of the fracture surface in a G-18/446 sample in planar and cross-section view. In this specimen fracture takes place within the chromate reaction layer. The yellow barium chromate is apparent on the washer surface of the failed sample in Figure 5(a), which is essentially stripped of the sealing glass. Figure 5(b) clearly shows fracture directly through the reaction zone layer.

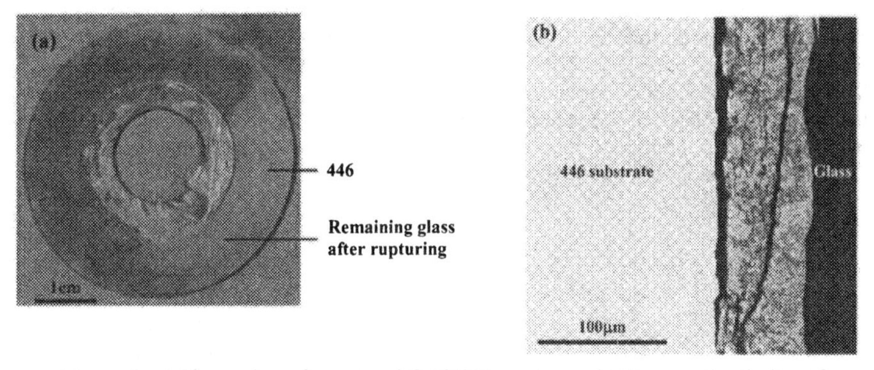

Figure 5 (a) Planar view of a ruptured G-18/446 specimen. (b) Cross-sectional view of the fractured portion of the same specimen. Note the complete delamination within the interfacial region between the glass and the metal substrate.

CONCLUSIONS

We have developed a screening test for glass pSOFC seals that allows us to examine the effects of a variety of materials, processing, and operating parameters on the hermeticity and strength of the resulting joint. Through the use of the rupture test, we have determined that alumina-forming ferritic steel substrates offer greater bond strength with the barium aluminosilicate based-glass employed here. The dominant factor in joint strength appears to be the composition and thickness of the reaction zone that forms in between the metal's oxide scale and the bulk glass. The barium chromate layer that develops on the chromia-forming steels exhibits poorer thermal expansion matching and tends to grow to a greater thickness than the corresponding celsian zone observed on the alumina-formers.

ACKNOWLEDGMENTS

The authors would like to thank Nathan Canfield for his preparation of the ceramic bilayer samples and Nat Saenz, Shelly Carlson, and Jim Coleman for their assistance in the metallographic analysis. This work was supported by the U. S. Department of Energy, Office of Fossil Energy, Advanced Research and Technology Development Program. The Pacific Northwest National Laboratory is operated by Battelle Memorial Institute for the United States Department of Energy (U.S. DOE) under Contract DE-AC06-76RLO 1830.

REFERENCES

1. B.C.H. Steele, A. Heinzel (2001) Materials for fuel-cell technologies, *Nature*, **414**(X) 345-52.
2. K. Eichler, G. Solow, P. Otschik, W. Schaffrath (1999) BAS ($BaO{\cdot}Al_2O_3{\cdot}SiO_2$) glasses for high temperature applications, *J. Eur. Cer. Soc.*, **19**(6-7) 1101-4.
3. S. P. Simner, J. W. Stevenson, K. D. Meinhardt, and N. L. Canfield in *Solid Oxide Fuel Cells VII*, edited by H. Yokokawa and S. C. Singhal, (The Electrochemical Society, Pennington, NJ, 2001) pp.1051.
4. K. D. Meinhardt, J. D. Vienna, T. R. Armstrong, and L. R. Peterson, PCT Int. Appl., WO01909059 (2001).
5. N. Lahl, L. Singheiser, K. Hilpert, K. Singh, and D. Bahadur: "Aluminosilicate Glass-Ceramics as Sealant in Solid Oxide Fuel Cells," in *Proceedings of the Sixth International Symposium on Solid Oxide Fuel Cells*, The Electrochemical Society, Pennington, NJ, 1999, pp. 1057-65.
6. Z. Yang, K. S. Weil, D. M. Paxton, and J. W. Stevenson: "Selection and Evalauation of Heat-Resistant Alloys for SOFC Interconnect Applications," *J. Electroch. Soc.*, 2003, *150*, pp. A1188-1201.
7. D. Bahat: "Kinetic Study on the Hexacelsian-Celsian Phase Transformation," *J. Mater. Sci.*, 1970, *5*, pp. 805-10.

FRACTURE IN Nb/Al_2O_3 PARTICULATE COMPOSITES

J. Matterson*, I. E. Reimanis* and J. Berger#
*Metallurgical and Materials Engineering Department
#Engineering Division
Colorado School of Mines
Golden, Colorado 80401
USA

ABSTRACT

Fully dense composites with the following volume percentages of Nb and Al_2O_3 were fabricated by hot pressing in a vacuum at 1550°C: 20/80, 40/60, 60/40 and 80/20. Specimens were machined into chevron notch bars and the fracture toughness was measured. The fracture toughness increases with volume percent Nb, rising from 5 MPa $m^{1/2}$ to 10 MPa $m^{1/2}$. No evidence of plastic deformation on the fracture surface was observed. The increase in toughness with Nb content was attributed to the presence of the inherently higher toughness of Nb.

INTRODUCTION

Joining is frequently accomplished by the presence of interlayers that are either ductile and relieve thermal residual stresses or provide a match in thermal expansion coefficient between the materials to be joined, thereby spreading the thermal residual stresses over a larger area. In the former case, it is necessary to know how to optimize the mechanical properties of the interlayer. In the latter case, compositionally graded interlayers may provide the most desirable thermal match between the parts [1,2]. In both cases, it is desirable to know the fracture behavior of the joint constituents. As part of a larger study concerning the fracture of graded interlayers [3-5], Nb/Al_2O_3 particulate composites were chosen so that elastic and plastic mismatch effects on fracture could be examined in isolation of complicating effects of thermal residual stress. In particular, Nb and Al_2O_3 exhibit nearly identical thermal expansion coefficients (~8×10^{-6}/°C [6]), leading to very small thermal residual stresses. Surprisingly, there are no reports in the literature on the mechanical properties of Nb/Al_2O_3 composites. The present paper determines the fracture toughness for four different composite layers of Nb/Al_2O_3 particulate composites.

EXPERIMENTAL METHODS

Powders of both Nb and Al_2O_3 (5μm and 0.48 μm average grain diameter respectively) were obtained to fabricate mono-composition composites in accordance with the desired compositional pairs found in Table I. Powders were weighed to produce the proper volume fractions and then mixed using a ball mill for eight hours with periodic agitation perpendicular to the roll direction to help prevent the formation of agglomerates. The powders were then sintered into pucks using a hot press from Thermal Technology Inc. (model number 610G-25T).

Table I. Compositions of the four Nb/Al_2O_3 composite specimens.

Volume Percent Nb	**Volume Percent Al_2O_3**
20	80
40	60
60	40
80	20

This hot press uses a graphite die and punch. In order to prevent the diffusion of carbon from the graphite die into the specimen, and thereby avoiding the formation of niobium carbide, the die material in contact with the sample was coated with inert boron nitride. The specimens were all sintered under the same hot press conditions of 1550 °C for a 70 min soak at $\approx 10^{-6}$ Torr, under an applied pressure of 45 MPa.

The resulting cylindrical pucks, approximately 25 mm in diameter and 25 mm in length, were machined into four point bend test chevron notch specimens[1] with geometries studied by Munz, Shannon, and Bubsey [7]. The geometry and orientation of these chevron notch specimens is depicted in Figure I where W is the width, B is the breadth, a_o is the distance from the edge of the sample to the crack initiation point and a_1 is the distance from the sample edge to the point where the notch face exits the side of the sample.

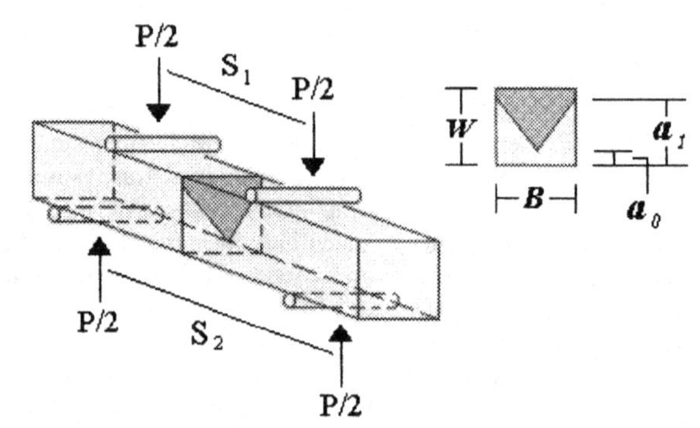

Figure I. Orientation and geometry of the chevron notch specimens.

There are many inherent benefits to using the four point bend chevron notch specimen geometry including the ability to calculate the plane strain fracture toughness using the maximum load without knowing a critical crack length [8]. In fact, through compliance testing, Munz, Bubsey, and Shannon [9], have determined that for materials with a flat crack growth resistance curve (R-curve), the plane strain fracture toughness, K_{IC}, is proportional to the maximum load and the dimensionless fracture toughness coefficient Y^*_m as seen in equation (1)[8].

$$K_{IC} = \frac{P_{max}}{B\sqrt{W}} Y^*_m \tag{1}$$

[1] Machined by BOMASS Inc., Boston, MA.

At fracture the load, P, will pass through a maxima and the dimensionless fracture toughness coefficient will pass through a minima [8]. B and W are geometric parameters defined in Figure I. Because equation (1) is limited to materials with flat R-curves, the accuracy will decrease with increasing ductility. In the present specimens strong R-curve behavior is not expected because both Al_2O_3 and Nb fracture in a brittle manner. To implement the four point bend test a stiff ceramic fixture of alumina was used in the configuration seen in Figure II. The inner and outer loading points, S_1 and S_2 respectively from Figure I, are fixed at 10 mm and 20 mm spans. The fixture is also fully articulated to compensate for any non-plane parallel surfaces. The load and displacement data was carefully acquired during the fracture of these specimens to determine the peak load for use in equation (1), and Y^* was determined from [8]. Four specimens were tested for each composition in Table I.

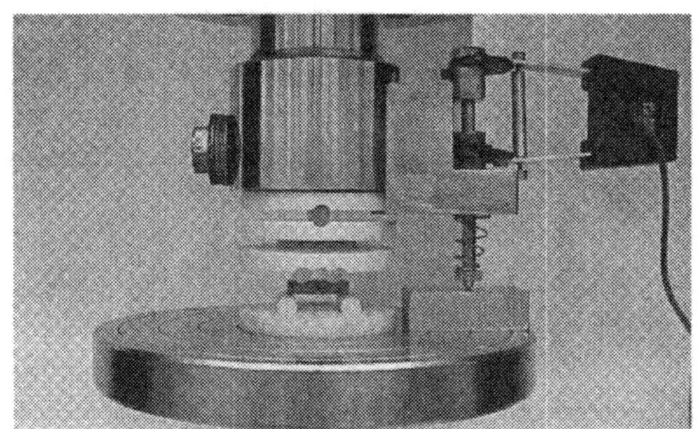

Figure II. Four point bending fixture used with chevron notch geometry. Displacement during the test was measured by the extensometer attached to the vertical pin on the right side of the figure.

RESULTS AND DISCUSSION

All specimens revealed a region of stable crack growth during the fracture experiment, thereby validating use of equation (1) in evaluating the fracture toughness. An example of the load-displacement curves for each of the four compositions is shown in Figure III. The fixture compliance has not been subtracted from these curves. The lack of obvious ranking of the slope of the curves with Nb volume percent may be due to changes in friction coefficient between the loading pins and the specimen for specimens of different composition.

The fracture toughness is shown as a function of volume percent Nb in Figure IV. Each point on the graph represents an average of the four specimens tested; the standard deviation is also given. Representative fracture surfaces for all compositions are shown in Figure V. An inhomogeneous microstructure is apparent in all specimens. Circular regions approximately 200 μm in diameter are Al_2O_3-rich, while surrounding regions contain more Nb. Figure VI reveals a higher magnification of the Al_2O_3-rich region in the 20 vol. % Nb specimen and the Nb-rich region in the 60 vol. % specimen.

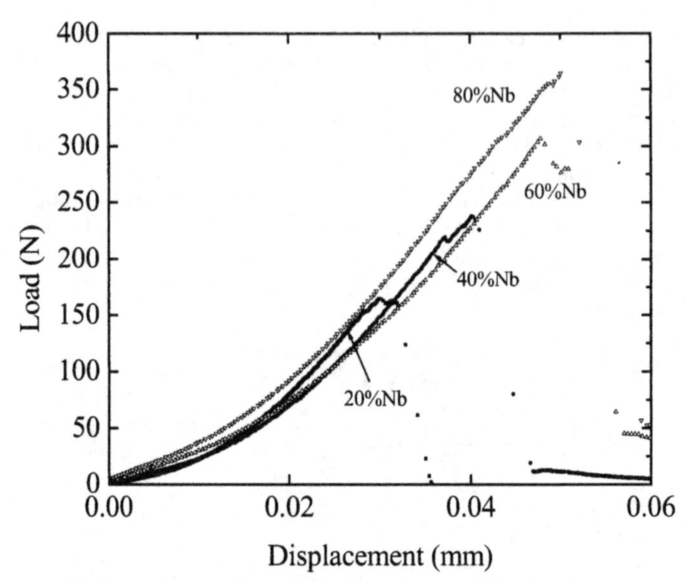

Figure III. Example raw load-displacement curves for the four compositions. Note that each curve shows evidence that stable crack growth was achieved, near the peak load.

Intergranular fracture is apparent in the Al_2O_3-rich regions, whereas transgranular cleavage fracture markings appear on the Nb regions. No evidence of plastic deformation in the Nb is visible. The features observed are identical for fracture surfaces of all compositions except that the relative amount of Al_2O_3 to Nb surfaces decreases with increasing Nb content. No evidence of any significant composite toughening mechanisms typically seen in metal/ceramic composites, such as pull-out, bridging, or crack blunting through plasticity was observed. This is not surprising since Nb does not exhibit much ductility [10], the interface between Nb and Al_2O_3 does not readily debond [11], and there is little or no thermal residual stress that might otherwise promote crack deflection. By considering that the energy released during fracture depends directly on the volume fraction of each phase present, a rule of mixtures law should accurately describe the variance of fracture energy with volume percent Nb, in the absence of any toughening mechanisms. Assuming linear elastic fracture, the data for fracture toughness in Figure 5 was used to obtain the fracture energy, Γ,:

$$\Gamma = \frac{\left(1-\upsilon^2\right)}{E} K_{IC}^2 \tag{2}$$

where ν and E are the Poisson's ratio and Young's elastic modulus for the particular Nb/Al_2O_3 composite. ν and E for Al_2O_3 are 0.22 and 375GPa [12], and for Nb are 0.38 and 103GPa [10].

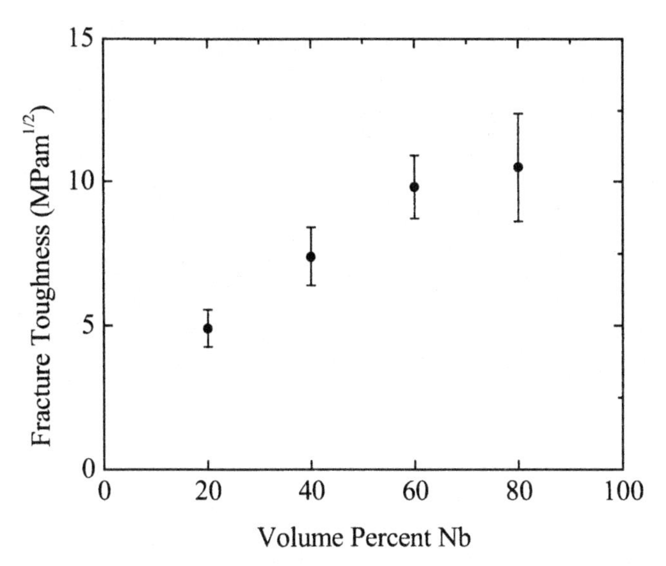

Figure IV. Fracture toughness measured by the chevron notch technique as a function of volume fraction Nb.

Using a linear rule of mixtures to determine the elastic constants and equation (2), the fracture energy is shown as a function of volume percent Nb in Figure VII. Also shown is the value of the fracture energy of pure Al_2O_3, based on an assumed value of 3 $MPam^{1/2}$ for the fracture toughness. The only known study in which the toughness of Nb was directly measured used the Charpy impact approach which provides a dynamic toughness [13]. The toughness reported was 37 $MPam^{1/2}$. Because of the sensitivity of the yield stress of Nb to impurity content, temperature and strain rate, a wide variety of toughness would be generally expected, and thus it was deemed inappropriate to compare pure Nb studied in the latter study with that in the present study. In any case, it is apparent from Figure VIII that a simple linear rule of mixtures description does not apply. Deviations from a rule of mixtures description may occur because of differences in the Nb constitutive properties between different compositions. Specifically, it would be expected that the oxygen partial pressure during hot pressing would be affected by the volume fraction of Al_2O_3 present. Nb present in the Al_2O_3-rich compositions may exhibit a more brittle fracture behavior and therefore contribute less to the overall toughness than Nb in the Nb-rich compositions.

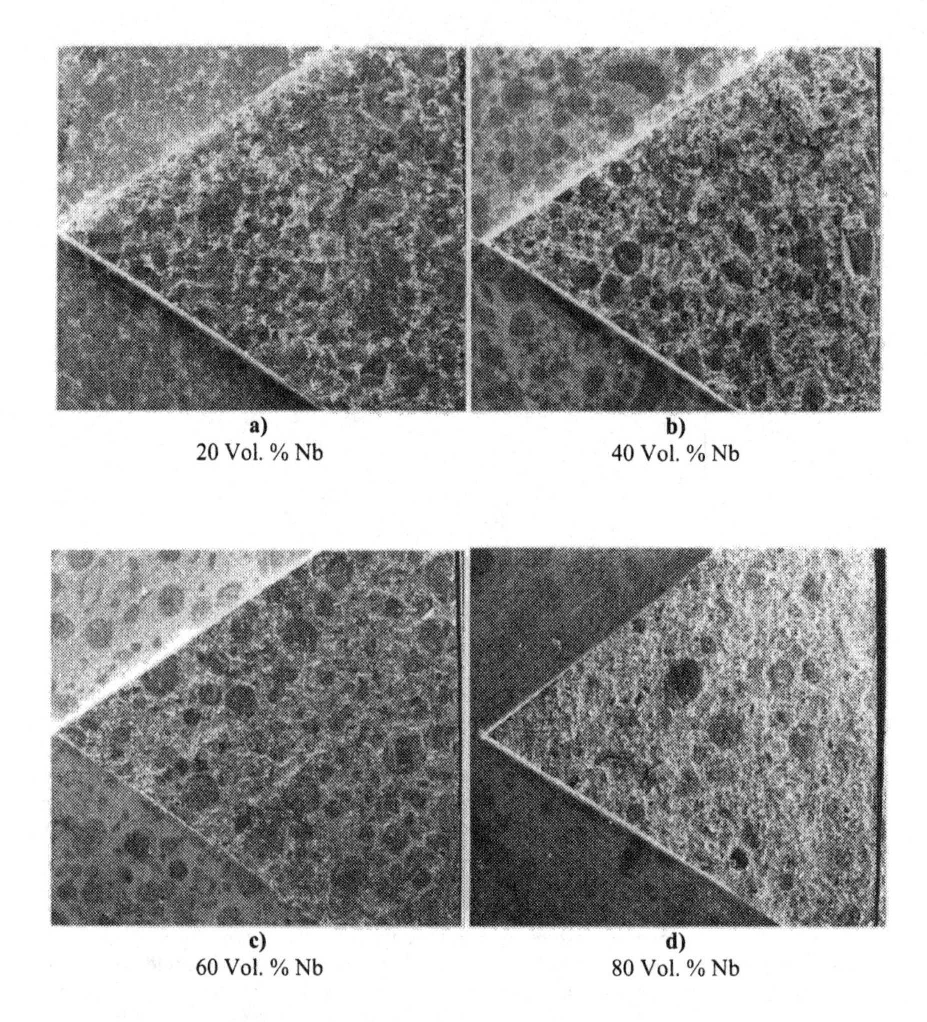

Figure V. Overview of chevron notch fracture surfaces. Light regions are Nb-rich; dark regions are Al_2O_3-rich.

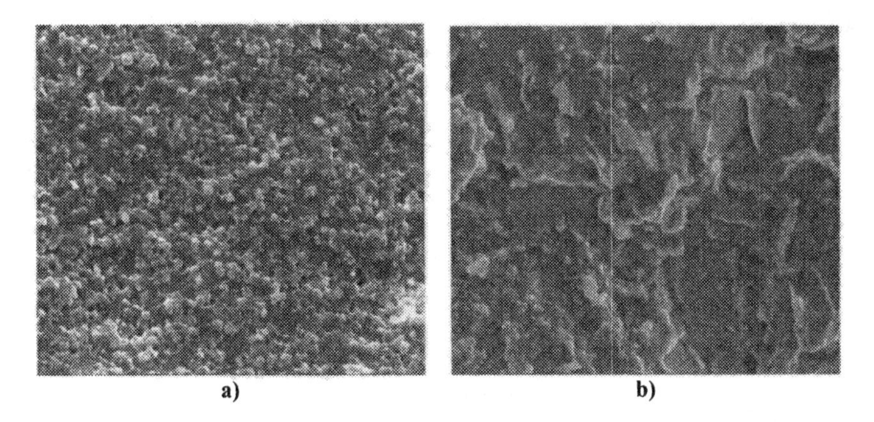

Figure VI. Higher magnification view of the fracture surface for a) 20 vol. % Nb in a Al_2O_3-rich regions and b) 80 vol. % Nb in a Nb-rich region. In the latter, fine grained Al_2O_3 is visible in the lower left part of the figure. Note the transgranular fracture in Nb, with no evidence of plastic deformation.

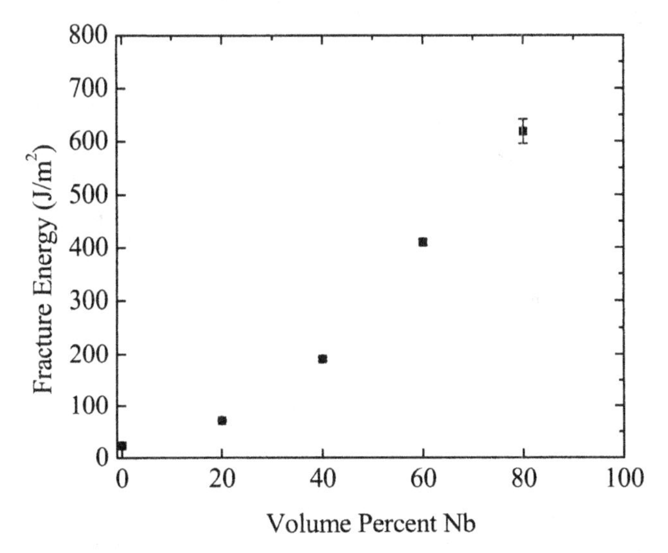

Figure VII. The fracture energy as a function of volume fraction Nb, calculated from equation (2) using data in Figure IV. Also included in the graph is a data point for pure Al_2O_3, assuming a fracture toughness of 3 $MPam^{1/2}$.

CONCLUSION

The fracture toughness of Nb/Al_2O_3 composites of compositions ranging from 20 vol. % Nb to 80 vol. % Nb was measured for the first time. It appears that no synergistic toughening mechanism operates, despite the presence of two distinct phases. The strong interfacial bond between Nb and Al_2O_3 likely does not promote debonding which is a prerequisite for1949chevy toughening by bridging. The lack of a linear dependence of the fracture energy with Nb content suggests that the degree of plasticity in Nb may change with different overall composite compositions. Future work will examine whether or not the Nb constitutive properties do indeed vary in the different compositions.

REFERENCES

[1]S. Suresh and A. Mortensen, Fundamentals of Functionally Graded Materials, IOM Communications Ltd. (1998).

[2]Y. Miyamoto, W. A. Kaysser, B. H. Rabin, A. Kawasaki, and R. G. Ford, Functionally Graded Materials: Design, Processing and Applications, Kluwer Academic Publishers, Boston (1999).

[3]J. Chapa-Cabrera and I. E. Reimanis. "Effects of Residual Stress and Geometry on Crack Kink Angles in Graded Composites", *International Journal of Engineering Fracture Mechanics,*Vol. 69, 14-16, pp 1667-1678, (2002).

[4]J. Chapa-Cabrera and I. Reimanis, "Crack Deflection in Compositionally Graded Cu-W Composites", *Philosophical Magazine A*, volume 82, number 17/18, 3393-3403 (2002).

[5]J. Stamile, J. Chapa-Cabrera and I. E. Reimanis, "Designing Joints with Graded Layers", *Ceramic Transactions* Vol. 138. Joining of Ceramic Materials. Eds. C. Lewinsohn, M. Singh, and R. Loehman. ISBN # 1-57498-153-6 (2003).

[6]Lee, Jong-Ki. "Electrochemical and Solid-State Letters,2." Substrate Effect on the Microstructure and Electrochemical Properties in the Deposition of a Thin Film $LiCoO_2$ Electrode 10 (1999): 512-515.

[7]D. Munz, J. Shannon and R. Bubsey "Fracture Toughness Calculations From maximum Load in Four Point Bend Tests of Chevron Notch Specimens" , *International Journal of Fracture,* Vol.16, R137-R141 (1980).

[8]A. Calomino, R. Bubsey "Compliance Measurements of the Chevron Notched Four Point Bend Specimen", NASA Lewis Research Center, to be published in J. Am. Ceerm. Soc.

[9]D. Munz, R. Bubsey and J.Shannon, *Journal of the American Ceramic Society* (1980) to appear.

[10]D. Douglass, F. Kunz, Columbium Metallurgy, Interscience Publishers, New York (1961)

[11]M. Ruhle, K. Burger, W. Mader and A. G. Evans, "Some Aspects of Structure and Mechanical Properties of Metal/Ceramic Bonded Systems" , *Fundamentals of Diffusion Bonding,* 43-68 (1980).

[12]CRC Materials Science and Engineering Handbook, 2nd Edition, Edited by J. F. Shackelford, W. Alexander and J. S. Park, CRC Press Boca Raton ISBN 0-8493-4250-3 (1994).

[13]Padhi, D. and J. J. Lewandowski, "Effects of Test Temperature and Grain Size on the Charpy Impact Toughness and Dynamic Toughness (K_{ID}) of Polycrystalline Niobium", Metall. and Mater. Trans. A. Vol. 34A, 976-978 (2003).

PRACTICAL ADHESION AND COHESION ASSESSMENTS OF Al_2O_3 (0.1μm) OXIDE LAYER ON TOP OF AlN SUBSTRATES BY MICROSCRATCH TECHNIQUE

Lotfi Chouanine, Masayuki Takano, Osamu Kamiya, and Makoto Nishida
Faculty of Engineering & Resource Science
Akita University
1-1 Tegata Gakuen-machi
Akita City 0100852, Japan

ABSTRACT

This paper presents experimental results for adhesion assessment of Al_2O_3 (0.1μm) oxide layer on different AlN substrate, electro-static chucks (ESC) for silicon plasma etching application. The AlN/Al_2O_3 substrates were scratched by a Rockwell C diamond indenter using a microscratch system operated in the progressive normal load mode. With increasing the load, the substrate/oxide layer generated stresses which, at a given load, resulted to cracking and chipping (cohesive failure), to delamination (adhesive failure), and to brittle fracture. A well-defined critical load (L_c) for each substrate/oxide layer was determined. The highest critical load (L_c = 0.45 N) and the lowest (L_c = 0.36 N) were found for the AlN substrates which have average grain sizes of 3.0 μm and 9.9 μm, respectively. For all samples, adhesion was found to be strong and the failure to be cohesive in the oxide layer. SEM observations of the scratch grooves showed that the cohesive failure mode appear to be in form of discontinuous chipping. The practical adhesion properties of the AlN/Al_2O_3 were best in the AlN substrate which has small grain's size, density and fracture toughness values.

INTRODUCTION

In recent years, considerable applied research has been devoted by the ceramic manufacturing community to the development of high performance ceramic components for semiconductor processing such as AlN/sapphire ESC, a device that holds down and often heats the silicon wafer during chemical vapor deposition or plasma etching.[1-2] Generally, the AlN-based ESC exploits the attractive features of AlN such as thermal uniformity, rapid heat up, easy temperature control, excellent plasma durability and less contamination.[3] However, due to its low toughness and poor machinability, damage free chucking and handling of the silicon wafers by the AlN ESC is not always achievable. In harsh vacuum environments, the fast high-temperature heating of the silicon wafer by the AlN ESC causes a sharp increase of the friction force at the interface between the AlN ESC and the wafer.[4-5] As a result, if the AlN ESC suffers considerable machining surface damages (SD) caused by polishing[6], intergranular cracking may cause disconnections of weak AlN particles, which in their turn may further contaminate and damage the chucked surface of the wafer.

Al_2O_3 oxide layers were produced on the AlN ESC substrates in order to improve their chucking performance in service. Accordingly, the objective of this tribological work was to optimize the properties of the AlN/Al_2O_3 substrates, so as to achieve desired properties and thereby reduce potential factors inducing failure. The influence of the grain size, density, fracture toughness and other properties of the AlN substrates on chipping (cohesive failure), delamination (adhesive failure), and brittle fracture of the substrate/oxide layer were investigated using micro scratch testing (MST) in combination with SEM and EDS examinations of the scratch tracks.[7-17]

MST is a comprehensive method of quantifying the scratch resistance, cohesion and adhesion of a wide range of bulk materials and ceramic films.[8-9, 17] MST involves generating a controlled scratch with a diamond tip on the sample under test.[7] The tip, a Rockwell C diamond, is drawn across the oxide layer under a progressive load.[9, 10-12] With increasing load, the AlN deformation generates stresses which, at a given load, result in permanent damage, such as chipping and cracking of the material or/and flacking of the oxide layer.[9] The smallest load leading to unacceptable damage or the critical load, L_c, can be precisely detected by means of an acoustic sensor attached to the indenter holder, frictional force, and by optical observations.

THEORETICAL CONSIDERATIONS

Figure 1 shows the forces acting on the diamond tip during the scratch test. F_n is the load applied on the tip of radius R, τ is the shearing force per unit area due to the deformation of the surface, and r is the radius of the circle of contact and therefore the half-width of the track left after the test. H is similar to a uniform hydrostatic pressure acting normal to the surface of indentation and can therefore be considered as the hardness of the AlN substrate. Assuming that the AlN substrate follows plastic deformation in scratch, the material surface hardness, H, can be related to the applied load and the radius of the contact circle, r:

$$H = \frac{F_n}{\pi . r^2} \quad (1)$$

The adhesion shear strength, τ, and the critical applied normal load, L_c, are related by[6]

$$\tau = H \tan\theta = \frac{L_c}{\pi a^2}\left[\frac{a}{\left(R^2 - a^2\right)^{1/2}}\right] \quad (2)$$

where L_c is the critical normal load, r is the contact radius, and R is the stylus radius. An accurate determination of the critical load sometimes is difficult. The tangential or friction force is measured during scratching to measure the critical load as well. The acoustic emission and the friction forces are very sensitive in determining critical load. The acoustic emission and friction force start to increase as soon as cracks begin (L_{c1}) to form perpendicular to the direction of the moving stylus. The work of adhesion, W, of the AlN/Al_2O_3 system is derived as a function of the normal critical load, L_{c2}, at which the delamination of the Al_2O_3 layer occurs [12, 14]:

$$W = \frac{\pi a^2}{2}\left(\frac{2EL_{c2}}{t}\right)^{\frac{1}{2}} \quad (3)$$

where, E is the Young's modulus of elasticity and t is the thickness of the Al_2O_3 oxide layer.

EXPERIMENTAL

Preparation of AlN/Al_2O_3 (0.1μm) Samples

Al_2O_3 oxide layers, 0.1 μm in thickness, were formed on AlN substrates by heating the material up to 1370 °C. The thickness of the oxide layer was kept unchanged for all samples and was measured by contact mode using a spherical abrasion (Calotest) method. We have tested 3 types of AlN substrates, two of them are dense technical grade materials and one includes additives such as yttrium. The material properties of these substrates are shown in Table I. Prior to oxidation, the substrates were polished to a surface roughness Ra = 10 nm using S_iO_2 slurry type (pH 03), thereafter sterilized with ethyl alcohol and cleaned with acetone in an ultrasonic bath for 15 min, and dried. AFM and SEM observations of the polished AlN substrates have revealed machining microcraks and chemical mechanical SSD caused by polishing.

Automatic Microscratch Testing

The scratches were performed using an MST apparatus operated in the progressive normal load mode, in the ranges of 0 to 10 N. A stylus with a Rockwell C type diamond of a tip radius of 10 µm was used. While a load was applied, the sample was moved at a constant speed, loading rates (0.2 N/min, and 3 N/min), sliding speeds (0.401 mm/min, and 1.788 mm/min), and applied load range (0-3 N, and 0-10 N). The scratch conditions are shown in Table II.

RESULTS AND DISCUSSION

Characterization of AlN Bulk Substrate without Al_2O_3 Oxide Layer on Top

The average measured critical loads of the scratches performed on the AlN bulk substrates (samples A, B and C) are given in Table III. An example of the results on AlN sample B (average grain size 5.9 µm) is shown in Fig. 2. The tangential force vs. the normal load is plotted in Fig. 2(a) between 0 and 3 N. The SEM micrographs of the scratch track corresponding to normal loads of 0.49 N and 3.0 N are shown in Figs. 2(b) and (c), respectively. Figs. 2a and b indicate early occurrence of several small crack grow events at shallow depths. The load corresponding to the nucleation and propagation of the first visible crack and chipping (low cohesion failure) is 0.27 N. The crack became more intensive (High cohesive failure) at critical normal load of 0.45 N as shown in Fig. 2b, where chipping and intergranular cracking are well seen around the scratch track. One explanation for early crack formation in the AlN bulk material is the pre-existence of microckracks induced by the polishing process. The pre-existing cracks need only propagation to cause cohesive failure of AlN substrate.

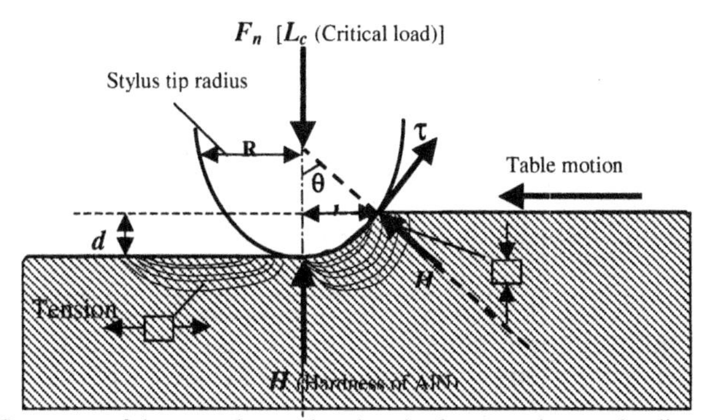

Fig. 1. Geometry of the scratch test showing the forces acting on the diamond tip

Table I. Material properties of ALN substrates

Mechanical & Thermal Properties		AlN substrates A (99.9%)	B (additives)	C (99.9%)
Density	(g/cm^3)	3.26	3.27	3.33
Strength	(MPa)	360	372	. 280
Hardness	(Hv)	1158	-	943
Young Modulus	(GPa)	-	-	321
Fracture Toughness	(Mpa.m$^{1/2}$)	2.2	2.8	3.8
Average grain size	(μm)	3.0	5.9	9.8
CTE	400 °C	3.8	4.6	4.6
PPM (10^{-6})	600 °C	4.5	5.0	5.1
Thermal conductivity	(W/m k)	90	94	190

Table II. Microscratch conditions used for all experiments

Experimental parameters	Progressive loading Scratch Conditions	
Scratch Mode	Low loading rate	High loading rate
Diamond Indenter	Rockwell C	Rockwell C
Tip Radius	10μm	10μm
AE sensitivity	9	9
Scratching length	6 mm	6 mm
Loading range	0 - 3N	0 - 10 N
Loading rate	0.2 N/min	3 N/min
Scratching speed	0.401 mm/min	1.788 mm/min

Table III. Average cohesive failure load L_{c1}, and adhesion failure load L_{c2} (three tests per sample) for the bulk AlN and for the AlN/Al_2O_3(0.1 μm) substrates between 0 and 3 N.

Substrate				Oxide layer		Average critical load (N)	
AlN	Purity %	grain size (μm)	Roughness Ra (nm)	Al_2O_3	Thickness t (μm)	Cohesive L_{c1} (N)	Adhesive L_{c2} (N)
A	99.9	3.0	50	No	-	0.253	
A[1]	99.9	3.0	50	Yes	0.1	0.45	0.64
B	Additives	5.9	50	No	-	0.213	
B[1]	Additives	5.9	50	Yes	0.1	0.37	0.59
C	99.9	9.8	50	No	-	0.243	
C[1]	99.9	9.8	50	Yes	0.1	0.36	0.58

Table IV. Average cohesive failure load L_{c1}, and adhesion failure load L_{c2} for the bulk AlN and for the AlN/Al_2O_3(0.1 μm) substrates between 0 and 10 N.

Substrate				Oxide layer		Average critical load (N)	
AlN	Purity %	grain size (μm)	Roughness Ra (nm)	Al_2O_3	Thickness t (μm)	Cohesive L_{c1} (N)	Adhesive L_{c2} (N)
A[1]	99.9	3.0	50	Yes	0.1	0.65	5
B[1]	Additives	5.9	50	Yes	0.1	0.46	3.18
C[1]	99.9	9.8	50	Yes	0.1	0.39	4.42

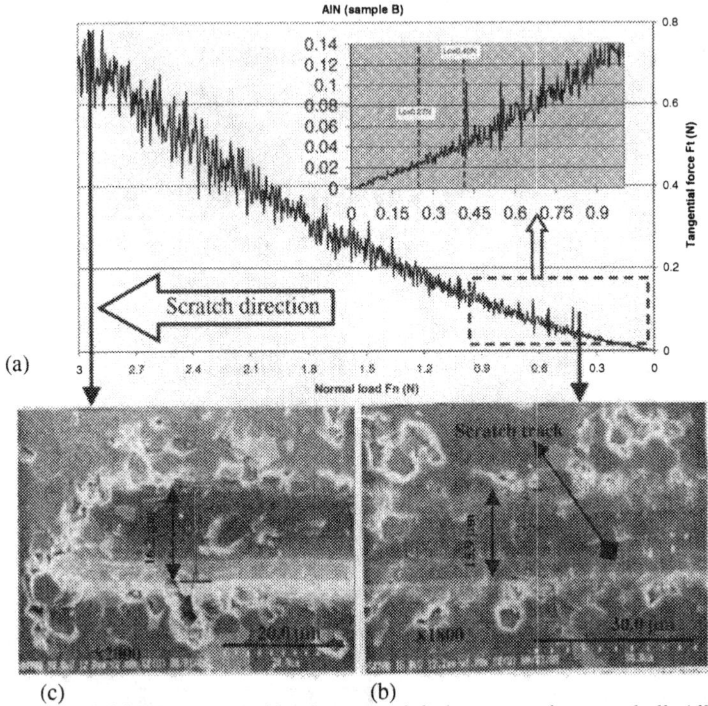

Fig. 2. (a) Tangential force vs. normal load measured during a scratch test on bulk AlN substrate type B, which includes additives (yttrium) and has an average grain size of 5.9 μm. (b)-(c) SEM images of the scratch track corresponding to normal loads of 0.49 N and 3 N, respectively.

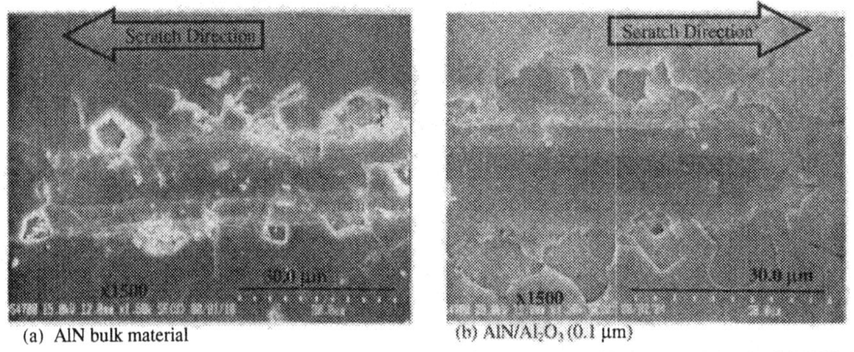

Fig. 3. SEM images of scratches corresponding to maximum applied normal load of 3.0 N. The average grain size of the AlN substrate is 9.8 μm. (a) Bulk AlN. (b) AlN/Al_2O_3 (0.1 μm) substrate.

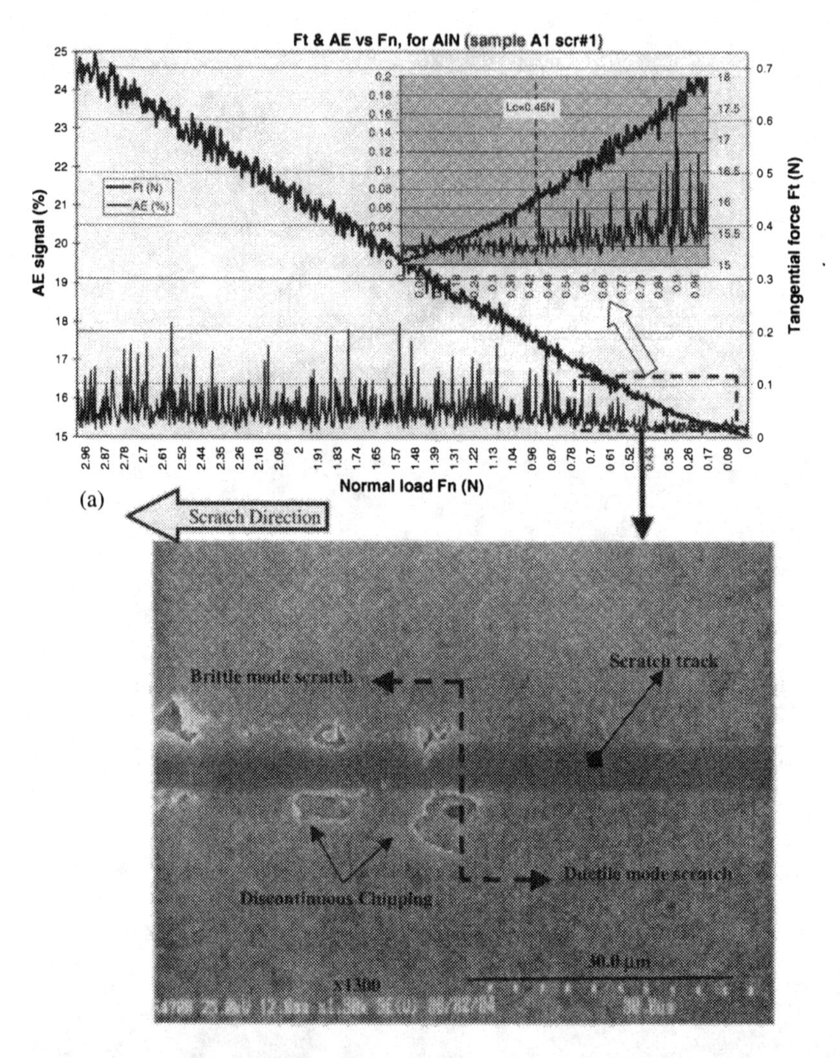

(b) Ductile to brittle transition mode scratch

Fig. 4. Scratch test on AlN/Al_2O_3(0.1 µm). The average grain size of the AlN substrate is 3.0 µm. (a) Tangential force and acoustic emission intensity vs. normal load between 0 and 3 N under a low loading rate, R = 0.2 N/min. (b) SEM image of the scratch corresponding to a critical normal load of L_c = 0.45 N.

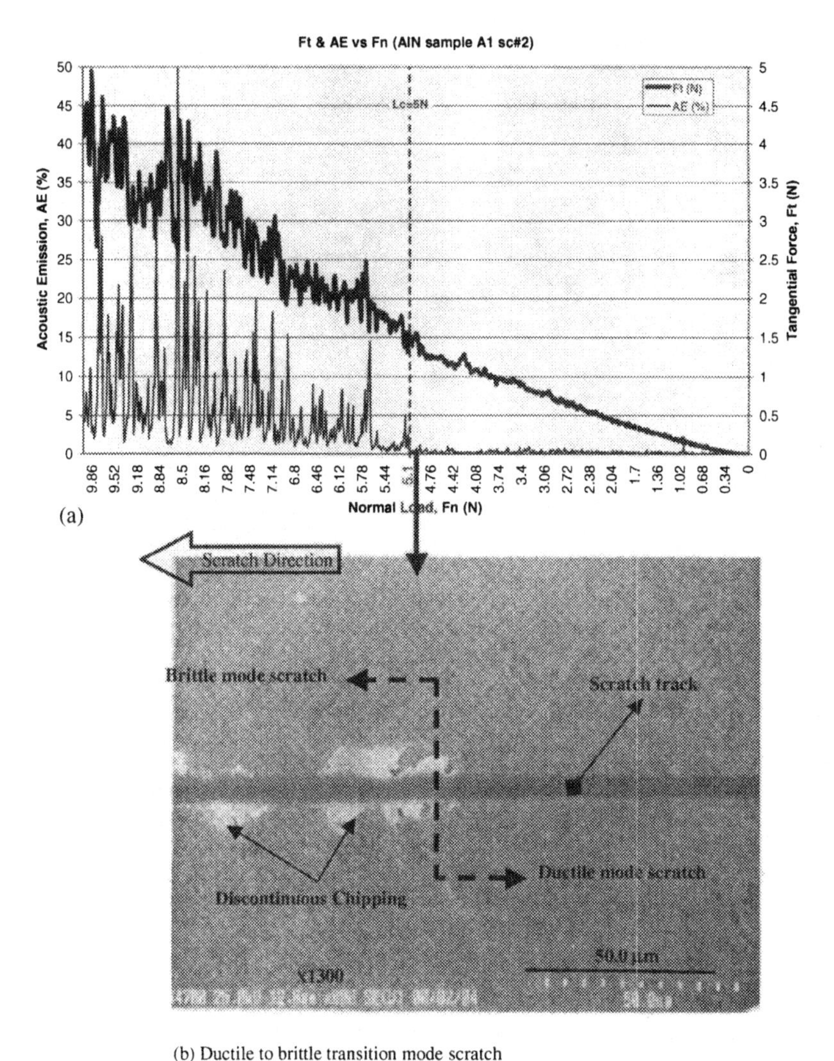

(b) Ductile to brittle transition mode scratch

Fig. 5. Scratch test on AlN/Al_2O_3(0.1 μm). The average grain size of the AlN substrate is 3.0 μm. (a) Tangential force and acoustic emission intensity vs. normal load between 0 and 10 N under a high loading rate, R = 3 N/min. (b) SEM image of the scratch corresponding to a normal load critical normal load of L_c = 5 N.

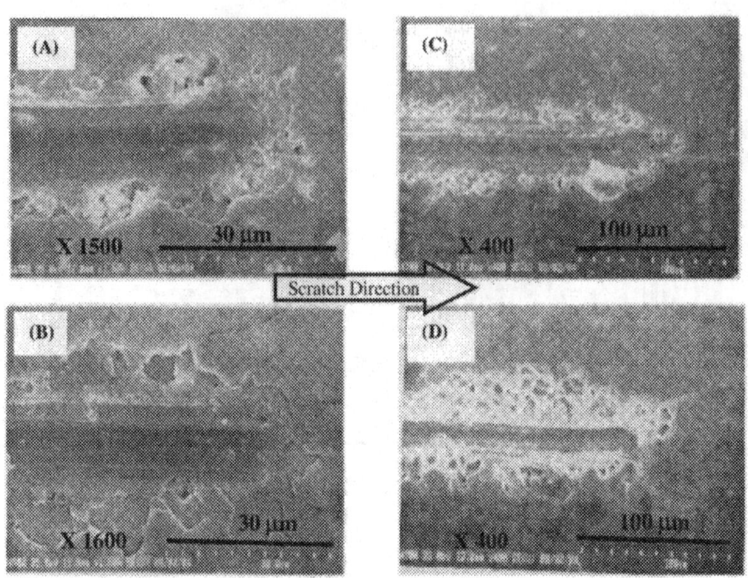

Fig. 6. (A-B) SEM images of scratches on AlN/Al_2O_3 (0.1µm) corresponding to maximum normal load, $\boldsymbol{F_n}$ = 3 N under a loading rate, R = 0.2 N/min. The average grain size of the AlN substrate is 3.0 µm (A), and 5. 9 µm (B). (C-D) SEM images of scratches on AlN/Al_2O_3 (0.1µm) corresponding to maximum normal load, $\boldsymbol{F_n}$ = 10 N under a loading rate of R = 3.0 N/min. The average grain size of the AlN substrate is 3.0 µm (C), and 5. 9 µm (D).

Characterization of AlN Bulk Substrate with Al_2O_3 Oxide Layer on Top

Figs. 3 (a) and (b) show the SEM micrographs of the scratches performed on bulk AlN sample B (average grain size 5.9 µm), and on AlN/Al_2O_3(0.1µm), respectively. Both images correspond to an applied maximum load of 3 N. In both cases, the cracking mechanism appears to be in form of discontinuous chipping. The results given in Table III show that, the measured average cohesive loads, L_{c1}, (nucleation and propagation of first crack and chipping) were higher for the AlN/Al_2O_3(0.1µm) specimens than that for the bulk AlN substrate.

For AlN/Al_2O_3(0.1µm), where the substrate is type A (average AlN grain size of 3 µm), the tangential force vs. the normal load is plotted in Fig. 4(a) between 0 and 3 N. Unlike in the case of the bulk AlN material (Fig. 2b), Fig. 4b indicates that cracks did not grow at small normal loads. The load corresponding to the nucleation and propagation of the first visible chipping is 0.45 N.

Influence of the Scratch Conditions on the Cohesion and Adhesion of AlN/Al_2O_3(0.1µm)

For AlN/Al_2O_3(0.1 µm), the substrate is type A (average AlN grain size of 3 µm), the AE signal intensity and the tangential force F_t are plotted in Fig. 4a vs. the normal progressive load between 0 and 3 N along a scratch length of 6 mm. The loading rate of 0.2 N/min, sliding speed of 0.401 mm/min were applied. The surface topography of the Al_2O_3 (0.1 µm) oxide layer is well illustrated by the SEM image of Fig. 4b. With increasing the normal load F_n, the oxide layer plastically deformed in a ductile mode. The first crack grow event occured in form of chipping at a critical load L_c = 0.45 N and corresponded to the first AE signal intensity. Fig. 4b combined with the EDX to identify exposure of the substrate clearly indicated no delamination of the oxide layer took place. Chipping initiated and propagated across the layer and traverse directly into the AlN substrate. With further advance of the indenter, another build-up of stored energy occurred within the oxide layer and the AlN substrate until a new crack initiated within the oxide layer and propagated across the substrate. This mechanism was repeated periodically all along the scratch.

Fig. 5a shows the AE signal intensity and the tangential force F_t plots vs. the normal progressive load between 0 and 10 N along a scratch length of 6 mm. The high loading rate of 3 N/min and a sliding speed of 1.788 mm/min were applied. The surface topography of the Al_2O_3 (0.1 μm) oxide layer is well illustrated by the SEM image of Fig. 5b. The fracture mechanism of the oxide layer, discontinuous chipping, was found very similar to the one described in Fig. 4. The load corresponding to the nucleation and propagation of the first chipping was found L_c = 5 N instead of L_c = 0.45 N for the case of low loading rate and sliding speed (Fig. 4).

Influence of the Substrate Properties on the Cohesion and Adhesion of AlN/Al_2O_3(0.1μm)

Fig. 6 show the SEM images of scratches on AlN/Al_2O_3 (0.1μm) corresponding to the applied maximum normal load, F_n = 3 N under a loading rate, R = 0.2 N/min, in case of the AlN substrate which has an average grain size of 3.0 μm (Fig. 6a), and in case of the substrate which ha an average grain size of 5. 9 μm (Fig. 6b). The size and area density of chipping around the scratch track seem to be not affected by the AlN grain size of the substrate. With increasing the scratching speed, the load range, and the loading rate, the fracture behavior did not change in form for all tested three AlN/Al_2O_3 (0.1μm) samples. The morphologies of the scratch tracks are well seen in these SEM images in Fig. 6. However, Fig. 6c shows also the substrate which has an average AlN grain size of 3.0 μm suffers less damage than the one which has an average AlN grain size of 5.9 μm (Fig. 6d). Accordingly, the critical load L_c at which cracks or chipping occur decreases with increasing the AlN's average grain size, the fracture toughness, and the material density. Although the resistance to cohesion was found to increase by 1.06 times (Table IV) for the AlN/Al_2O_3(0.1μm) system comparing with that for the AlN bulk material, it should be better not to use it in siliconductor processing components because the Al_2O_3 oxide layer can lead to contamination of the SW during plasma etching.

CONCLUSIONS

In order to enhance the chemical, physical and thermal properties of the AlN ESC and to make it indispensable for a successful performance in service, we generated on the surface an Al_2O_3 oxide layer, 0.1 μm in thickness. Thereafter, the main objective of this tribological work was to optimize the AlN/Al_2O_3 properties to achieve desired properties and a reduction of factors inducing failure. The results showed that automated microscratch testing proved to be a simple, expedient, and powerful technique for comparatively characterizing the cohesive and adhesive failure loads for both AlN bulk materials and for the AlN/Al_2O_3 system.

The cohesion and adhesion resistance of AlN/Al_2O_3 was found to be stronger than that measured for the bulk AlN substrate, to increase with increasing the loading rate and the scratching speed, and to decrease with increasing the average grain size of the AlN substrate. Unlike for the bulk AlN substrate, where the scratch results showed series of crack grow events, which occurred at different small normal loads, the cracking mechanism of AlN/Al_2O_3 appeared in form of discontinuous chipping with clear ductile to brittle transition scratch mode. Finally, the

REFERENCES

[1]P. R. Choudhury, "Handbook of Microlithography, Micromachining, and Microfabrication," Vol. 1, pp. 11-474, The Society of Photo-Optical Instrumentation Engineers, Washington, 1997.

[2]P. R. Choudhury, "Handbook of Microlithography, Micromachining, and Microfabrication," Vol 2, pp. 3-298, The Society of Photo-Optical Instrumentation Engineers, Washington, 1997.

[3]D. W. Richerdson, "Modern Ceramic Engineering: Properties, Processing, and Use in Design," 2nd Ed., p 374, Marcel Dekker, New York, 1992.

[4]T. Tsukizoe, T. Hisakado, "On the Mechanism of Contact Between Metal Surfaces: part2 - The Real Contact Area and the Number of Contact Points," ASME Jour. Lubrification Technology, **90F** 81-88 (1968).

[5]D. J. Whithouse, "Handbook of Surface Metrology," p. 792, The Institute of Physics, 1994.

[6]J. C. Lambropoulos, S. D. Jacobs, B. Gillman, F. Yang, and J. Ruckman, "Subsurface Damage in Microgrinding Optical Glasses," pp 469-74 in Proceedings of the 5th Inter. Conf. on Advances in the Fusion and Processing of Glass, Ceramic Transactions, Vol.82, 1998.

[7]P. Benjamin, and C. Weaver, "Measurement of Adhesion of Thin Films," Proc. R. Soc. London, **A254** 63-76 (1960).

[8]K. L. Mittal, "Adhesion Measurement of Thin Films, Thick Films and Bulk Coatings," pp.5-107, ASTM Philadelphia, 1978.

[9]K. L. Mittal, "Adhesion Measurement of Films and Coatings," VSP BV, Netherlands, 1995.

[10]A. J. Perry, "The Adhesion of Chemically Vapour-Deposited Hard Coatings on Steel-The Scratch Test," Thin Solid Films, **78** 77-93 (1981).

[11]A. J. Perry, "Scratch Adhesion Testing of Hard Coatings," Thin Solid Films, **197** 167-180 (1983).

[12]P. A. Steinmann, Y. Tardy, and H. E. Hintermann, "Adhesion Testing by the Scratch Test Method. The Influence of the Intrinsic and Extrinsic Parameters on the Critical Load," Thin Solid Films, **154** 333-349 (1987).

[13]P. J. Burnet, and D. S. Rickerby, "The Relationship Between Hardness and Scratch Adhesion," Thin Solid Films, **154** 403-416 (1987).

[14]Julia-Schmutz, and H. E Hintermann, "Microscratch Testing to Characterize the Adhesion of Thin Layers," Surface Coatings Technol., **48** 1 (1991).

[15]S. J. Bull SJ, and D. S. Rickerby, "New Developments in the Modeling of the Hardness and Scratch Adhesion of Thin Films," Surf. Coatings Technol, **42**, 149-164 (1990).

[16]T. W. Wu, "Microscratch Test and Load Relaxation Tests for Ultra-Thin-Films," J. Mater. Res., **6** 407-426 (1991).

[17]L. Chouanine, "Nanotribolgy of Thin Films in a Lithium Niobate Optical Waveguide Modulator $LiNbO_3$-Ti(20nm)-Pt(10nm)-Au(100nm)-Au(10μm) Using Depth-Sensing Nanoindentation and Scratching Technique," Optical Eng., **40** [8] 1709-1716 (2001).

WETTING AND MECHANICAL CHARACTERISTICS OF THE REACTIVE AIR BRAZE FOR YTTRIA-STABILIZED ZIRCONIA (YSZ) JOINING

J. Y. Kim, K. S. Weil, and J.S. Hardy
Pacific Northwest National Laboratory
902 Battelle Blvd.
Richland, WA 99352

ABSTRACT

Reactive air brazing (RAB) technique was developed as an effective alternative for the joining of complicated ceramic parts. It was found that additions of CuO to silver exhibit a tremendous effect on both the wettability and joint strength characteristics of the subsequent braze relative to polycrystalline yttria-stabilized zirconia (YSZ) substrates. The wettability of Ag-CuO braze on YSZ was substantially improved with CuO contents in the observed range between 1 an 8 mol%. The corresponding bend strength of the brazed YSZ joints appearred to moderately increase with CuO content. The addition of a small amount of titanium also helped improve wettability, especially at the low CuO content. However, no corresponding substantial improvement was observed in the titanium-added braze.

1. INTRODUCTION

As the operating temperature of advanced high-temperature electrochemical devices continues to be pushed upward based on thermal efficiency considerations, there is an increasing need to develop joining techniqes suitable for these applications, since joining is one of the economical ways to fabricate a large complex structure. Although a number of ceramic joining techniques currently exist, as is inherent with most materials technologies, each requires some form of trade-off or exhibits some penalty in terms of joint properties, easiness of processing, and/or cost [1-4]. Recently, an alternative joining technique, referred to as "reactive air brazing" (RAB), was developed [5, 6]. This braze consists of at least two components, a noble metal matrix and an oxide compound that is at least partially dissolved in a noble metal matrix. The oxide compound in the braze reactively modifies ceramic faying surfaces such that the newly formed surface is readily wetted by the remaining molten filler material. In many respects, this concept is similar to active metal brazing [4], except that the joining operation can be conducted in air and the final joint should be resistant to oxidation at moderate-to-high temperatures.

One system that appears to be readily suited for RAB is the Ag-CuO system. Meier et al. [7] demonstrated in a series of sessile drop experiments conducted in inert atmosphere that the contact angle between silver and alumina is greatly reduced by small additions of CuO. Schüler et al. [8] recognized that the CuO-Ag system could be exploited to bond ceramics in air, reporting their findings on a 1 mol% CuO-Ag braze composition used to join alumina substrates for power semiconductor packaging.

In this study, various Ag-CuO compositions in the range between 1 and 8 mol% of CuO were used as a starting point for developing a reactive air braze for yttria-stabilized zirconia (YSZ) whicha are commonly used as electrolyte in high-temperature electrochemical devices. The effect of small amount of titanium addition was also investigated in the same range of CuO content. Three types of experiments were performed in this investigation: (1) sessile drop measurements to determine the wetting behavior of the different Ag-CuO braze compositions on polycrystalline YSZ, (2) microstructural analysis on the braze/YSZ interfacial regions to determine the nature of wetting, and (3) four-point bend testing of as-brazed joints to measure joint strength at room temperature.

2. EXPERIMENTAL PROCEDURE

2.1 Materials

As listed in Table I, eight different braze compositions were selected for this study. These compositions were formulated by ball-milling the appropriate amounts of copper powder (99%, Alfa Aesar), silver powder (99.9%, Alfa Aesar) and/or titanium hydride (98%, Aldrich). The copper and titanium hydride oxidize *in-situ* as the braze is heated for joining, forming CuO and TiO_2. Zirconia discs, measuring ~3 cm in diameter and ~0.5 cm in thickness, were used as substrates in wetting experiments. To make these discs, 8 mol% yttria-doped zirconia powder (Tosoh) was isostatically cold-pressed at 150 MPa and then cold-pressed discs were fired at 1450°C for 1 h. The discs were polished on one face to a 10 μm finish. For the wetting studies, the mixed powder of each braze composition was cold-pressed into pellets, measuring approximately 7 mm in diameter and 10 mm high. The pellet density was ~ 65% of the theoretical density based on a rule of mixtures calculation for the dry starting materials

Table I. Braze compositions employed in this study

Braze ID	Ag (mol%)	Cu (mol%)	TiH_2 (mol%)
CA01	99	1	-
CA02	98	2	-
CA04	96	4	-
CA08	92	8	-
CA01-T	98.5	1	0.5
CA02-T	97.5	2	0.5
CA04-T	95.5	4	0.5
CA08-T	91.5	8	0.5

Polycrystalline ziconia plates containing 8 mol% yttria (ZDY-8, CoorsTek.) were used for bending tests. YSZ plates measuring 100 mm x 25 mm x 4 mm were joined along the long edge using a braze paste to form a 100 mm x 50 mm x 4 mm plate. Braze pastes were prepared by mixing a polymer binder (B75717, Ferro Corp.) and dry powder in a 1:1 weight ratio. After firing at 1000°C for 1 h, bend bars (50 mm x 4 mm x 3 mm) were cut from the plates

2.2 Testing and Characterization

Wetting experiments were performed in a static air box furnace, furnished with a large quartz window on the front door through which the heated specimen could be observed. A high speed video camera equipped with a zoom lens was used to record the wetting specimen during an entire heating cycle. Each braze pellet was placed on the polished side of a YSZ disc and heated at 30°C/min to 900°C, at which the heating rate was reduced to 10°C/min for the subsequent heat treatment. The furnace temperature was raised to 1000°C, where the temperature was held for 15 min. The temperature was also held at 1050 and 1100°C for 15 min for measuring contact angles. By holding temperature for 15 min, the contact angle between the braze and YSZ was allowed to stabilize for measurement at each temperature. Using VideoStudio6™ (Ulead

Systems, Inc.) software, selected frames from the videotape were converted to computer images, from which the wetting angle between the braze and YSZ substrate were measured and correlated with the temperature log for the heating run.

Joining samples were prepared by spreading the braze paste on the faying surface of each YSZ plates. Spring steel side clips and appropriately positioned refractory brick were used to keep the arrangement from slipping or toppling during a heating cycle. Specimens were joined incorporating a following heating schedule; heat in static air at 3°C/min to 1000°C, hold at 1000°C for 1 h and cool to room temperature at 5°C/min. Once joined, each sample was machined into rectangular bend bars, measuring 4 mm x 3 mm x 50 mm with the joint midway along their lengths. The edges of the side which are to be placed under tension during bending (bottom on bending test) were chamfered to remove machining flaws that could intiate premature failure. Strength testing was conducted by 4-point bending. The spans between the inner and outer contact points were 20 and 40 mm, respectively, and testing was performed at a displacement rate of 0.5 mm/min. Bend strengths were calculated from the load at failure using standard relationships derived for monolithic elastic materials [9]:

$$\sigma_F = \frac{3P \cdot L}{4b \cdot d^2}$$

Where P is the applied load, L is the length of the outer span, and b and d are the respective width and height of the specimen. Five specimens, each cut from the same plate, were used to determine the average room temperature joint strength for a given braze composition.

Microstructural analysis of the joints was performed on polished cross-sectioned samples of sessile drop experiments, using a JEOL JSM-5900LV scanning electron microscope (SEM). The SEM is equipped with an Oxford energy dispersive X-ray analysis (EDX) system, which employs a windowless detector for quantitative detection of elements. To avoid electrical charging of the samples in the SEM, they were carbon coated and grounded. Elemental profiles were also recorded across joint interfaces in the line-scan mode.

3. RESULTS AND DISCUSSION

3.1 Sessile Drop Experiments

The contact angles of eight braze compositions on the polished polycrystalline YSZ substrates were plotted in Figure 1. The fifteen minute hold time used in the sessile drop measurements appeared to be long enough for interfacial equilibrium to be established. In all cases, the contact angle reached its stable value within 5 min. Figure 1a shows a plot of contact angle as a function of braze composition for the two hold temperatures (1000°C and 1100°C). The contact angle displays a monotonic decrease (increase in wettability) with increasing CuO content. When a small amount of titanium hydroxide (0.5 mol%) was added, the wettability was further improved at given CuO content. The effect of titanium addition seems to decrease at a high CuO content, at which the contact angle is already small. On the other hand, the effect of temperature on contact angle is marginal (Fig. 1b), suggesting that the wetting phenomena, which take place on the braze/YSZ interface, are rapid and essentially completed at 1000°C. Thus, it is relatively unaffected by an increase in temperature, but may be hindered from reaching their maximum effect by the lack of reactant, i.e. too little CuO.

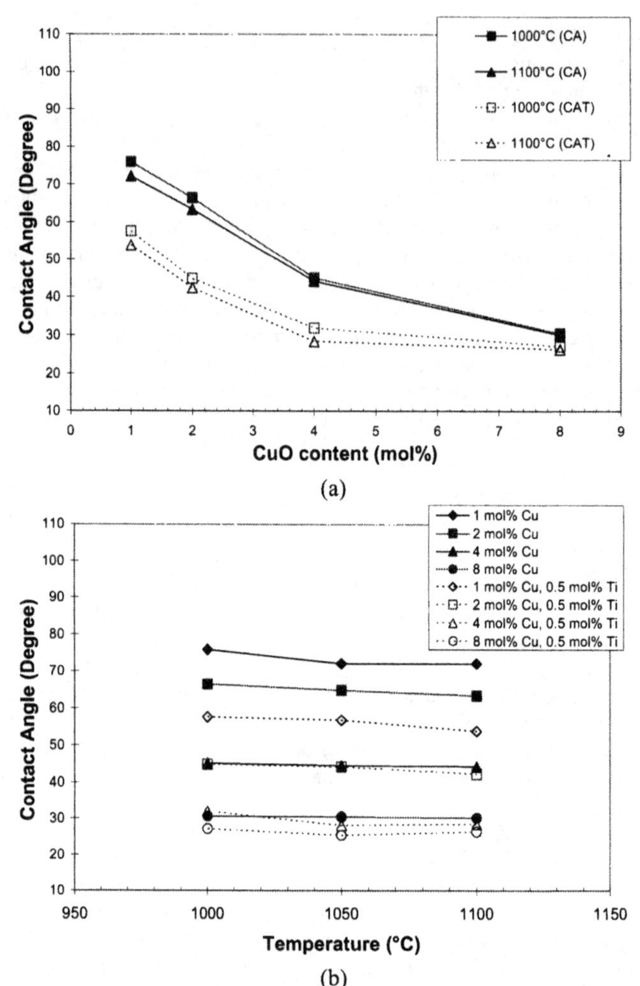

Figure 1 Contact angle as a function of (a) composition and (b) temperature. "CA" represents mixtures of copper and silver, while "CAT" represents mixtures of copper, silver and titanium.

3.2 Microstructural Analysis

Back scattered electron images collected on cross-section of sessile drop experiment samples are shown in Figures 2 and 3. Samples joined using the Ag-CuO braze ("CA") revealed presence of CuO particles at the interface between the braze and YSZ substrate as well as in the Ag matrix. The amount of CuO phase at the braze/YSZ interface increases with increasing CuO content in the braze, indicating that it is a preferential site for CuO precipitation from the Ag-CuO melt.

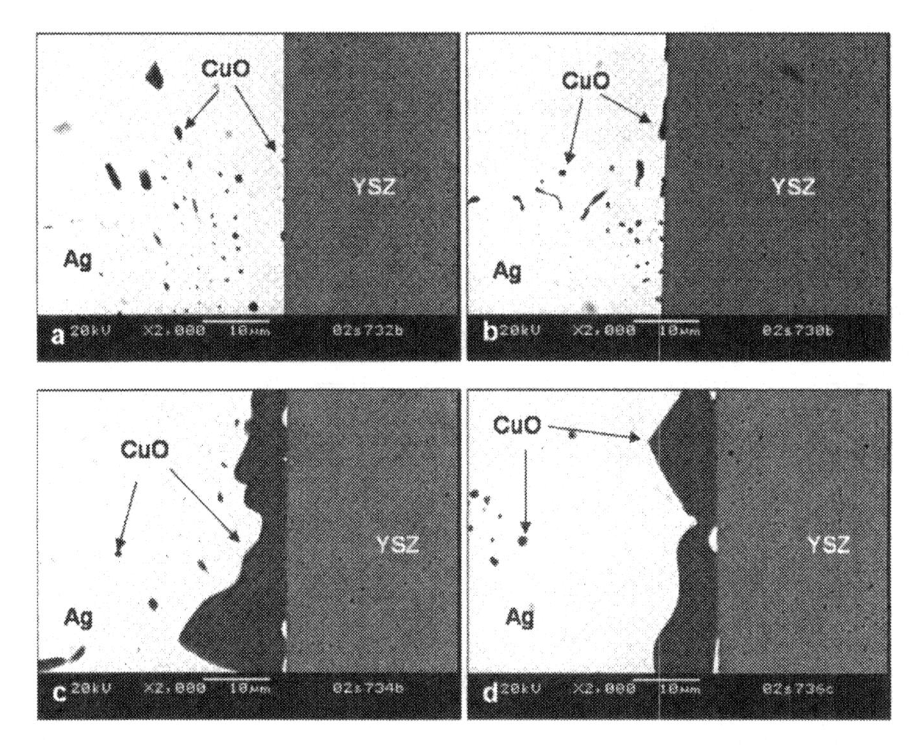

Figure 2 Cross-sectional SEM micrographs of the interface between the YSZ substrate and Ag-CuO braze: (a) CA01, (b) CA02, (c) CA04, and (d) CA08.

This result implies that there is affinity between CuO and YSZ, explaining why wettability improves with CuO content. Despite the affinity, no reacted phase was observed at the interface unlike CuO/alumina interaction which forms $CuAlO_2$ at the interface.

Microstructures of titanium-added braze joints ("CAT" samples) are shown in Figure 3. Similar to the "CA" braze, CuO was also preferentially placed near the braze/YSZ interface. However, the small amount of titanium (0.5 mol%) caused significant microstructural changes at the interface, forming a titanium zirconate phase at the interface. This result indicates that the improvement in wettability accomplished by titanium addition is related to the formation of an interfacial titanium zirconate phase. This titanium zirconate phase seems to form via cross-diffusion of both species, leaving a reacted phase on both sides of the original interface between the braze and YSZ substrate. Thus, all the titanium-added braze compositions except the CA01-T braze formed a continuous interfacial titanium zirconate phase on the YSZ substrate (~2 µm thick) and titanium zirconate particles on the braze side, while the CA01-T (Figure 3(a)) exhibited only interfacial titanium zirconate particles. This cross-diffusion of Zr and Ti also created voids due to the Kirkendall effect between particulate and continuous titanium zirconate phases, which were filled with the Ag-CuO melt.

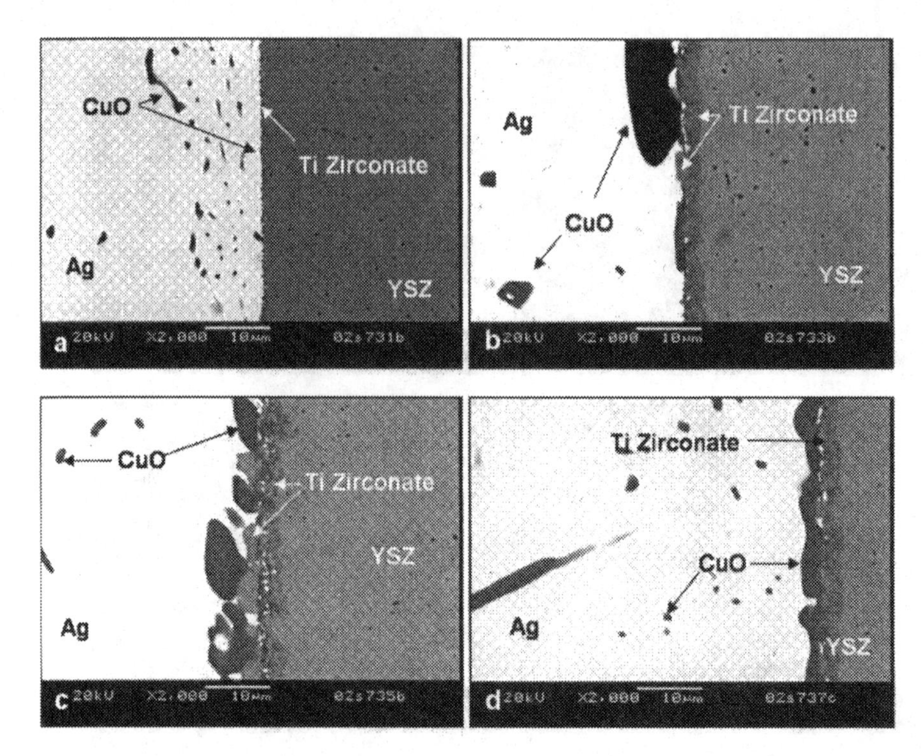

Figure 3 Cross-sectional SEM micrographs of the interface between the YSZ substrate and titanium-added Ag-CuO braze: (a) CA01-T, (b) CA02-T, (c) CA04-T, and (d) CA08-T.

The EDS line profile collected on the CA04-T sample is shown in Figure 4. The EDS analysis clearly reveals Kirkendall voids filled with Ag and CuO as well as the presence of particulate and continuous titanium zirconate phases. CuO particles were also found near the interface, but on top of particulate titanium zirconate. This result implies that the titanium zirconate phase formed before CuO phase was precipitated from Ag-CuO melt. The fact that voids were filled with Ag (and some CuO) also implies that the formation of titanium zirconate and Kirkendall voids probably occurred when the Ag-CuO was in a molten state.

3.3 Mechanical Property of RAB specimen

The 4-point bend strength of each joint, as measured at room temperature, is plotted along with contact angle as a function of CuO content in Figure 5. The bend strength of samples joined with the Ag-CuO braze ("CA" samples) reveals a moderate increase with CuO content in the range of 1 ~ 8 mol% CuO, mainly due to the improvement in wettability. The maximum strength of 111 MPa was achieved with the joint containing 8 mol% CuO, which is 56% of monolithic YSZ (197 MPa). In the case of titanium-added braze ("CAT" samples), bend strength increases with CuO content up to 6 mol% and decreases at 8 mol%.

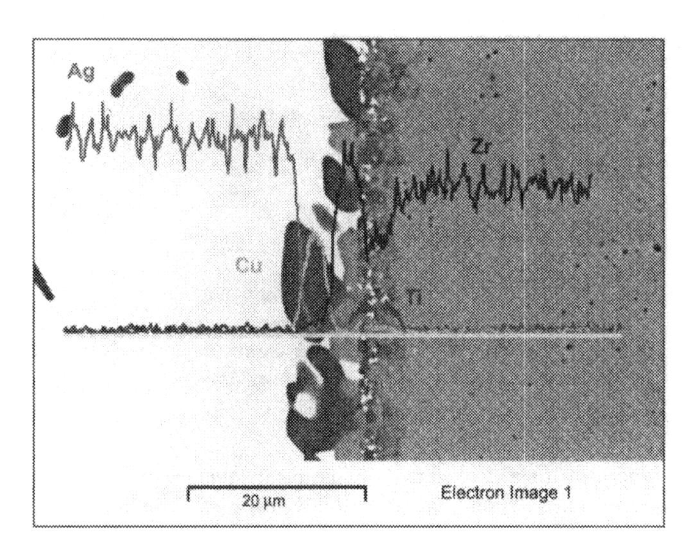

Figure 4 EDX analysis (line profiles) of Ag, Cu, Zr, and Ti obtained on the CA04-T braze.

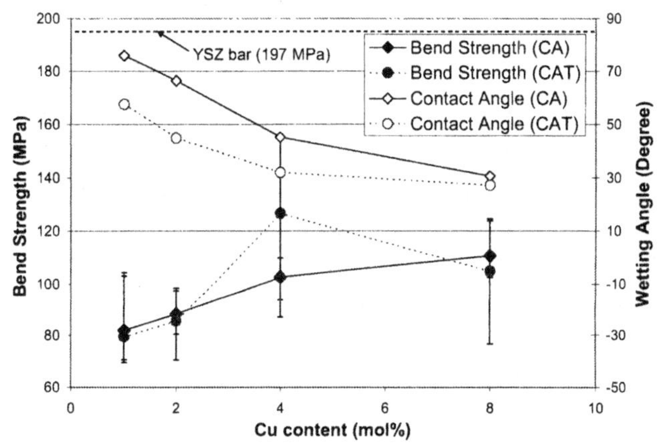

Figure 5 Room temperature 4-point bend strength and contact angle as a function of Cu content. The contact angles shown in the figure are values collected at 1000°C.

Although the contact angle is smallest at 8 mol%, it appears that a continuous phase of CuO along the interface plays a deleterious role in the strength of the joint (refer to Figure 3d). It should be also noted that the addition of titanium does not seem to improve bend strength of joints despite its effect on wettability. Only noticeable improvement in strength due to the titanium addition appeared on the specimen containing 4 mol% CuO (~20% increase in bend strength).

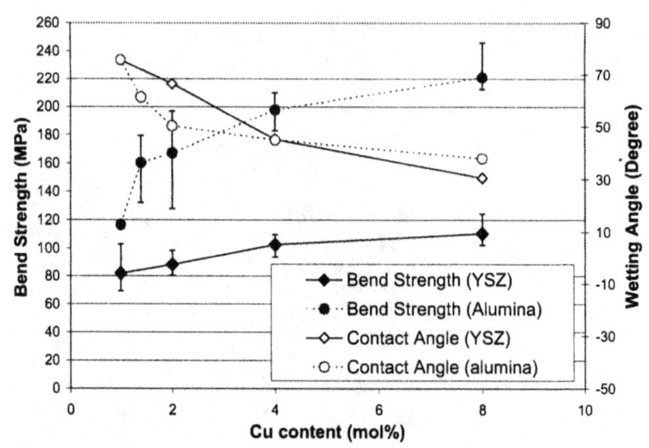

Figure 6 Comparison of room temperature 4-point bend strength and contact angle between alumina [6] and YSZ joints brazed at 1000°C using Ag-CuO braze.

Effects of CuO content on the strength of YSZ joints are compared with that of alumina joints in Figure 6. Although a general trend in wettability as a function of CuO content is similar in both cases, alumina joints revealed overall higher strength and more drastic increase in bed strength with CuO content in comparison to YSZ. The important difference in microstructure is that CuO in the braze partially diffuses into alumina substrates and sometimes forms a reacted phase such as $CuAlO_2$ [6], while YSZ does not reveal any reaction with CuO. Therefore, the fracture in YSZ joints propagated mostly along the braze/YSZ interface, while the fracture in alumina joints propagated through the Ag-CuO braze filler when CuO content in the braze is less than 8 mol% [6]. Thus, although the addition of CuO improves the wettability of the braze on YSZ due to its affinity with YSZ and hence bend strength to some extent, the improvement in the bend strength is limited for YSZ unlike alumina in which CuO particles works as pinning points holding the braze/alumina interface.

CONCLUSIONS

Reactive air brazing (RAB) using Ag-CuO brazes was investigated as a potential technique applied for high-temperature electrochemical applications. It was found that the addition of copper oxide to silver significantly improved the wettability of the braze on YSZ. A small amount of titanium addition also improved the wettability considerably, especially for the braze with low CuO contents. The bend strength of YSZ joints also revealed increase in bend strength mainly due to improvement in wettability. The maximum strength of 127 MPa was achieved with the braze containing 4 mol% of CuO and 0.5mol% of TiO_2, which was 64% of the bend strength of monolithic YSZ used in this study. However, the effect of CuO addition on the strength of YSZ joints is less drastic in comparison to alumina, possibly due to the limited interaction between CuO and YSZ.

ACKNOWLEDGMENT

The authors would like to thank Nat Saenz, Shelly Carlson and Jim Coleman for their assistance in preparation of metallographic samples and SEM analysis work. This work was supported by the U. S. Department of Energy, Office of Fossil Energy, Advanced Research and Technology Development Program.

REFERENCES

[1]K. Eichler, G. Solow, P. Otschik, and W. Schaffrath, "Degradation Effects at Sealing Glasses for the SOFC"; pp. 899-906 in *Proceedings of the Fourth European Solid Oxide Fuel Cell Forum*. Edited by A.J. McEvoy. Oberrohrdorf, Switzerland (2000).

[2]M. Singh, T. Shibayama, T. Hinoki, M. Ando, Y. Katoh, and A. Kohyama, "Joining of Silicon Carbide Composites for Fusion Energy Applications," *Journal of Nuclear Materials*, **283-287B** 1258-61 (2000).

[3]E. Pippel, J. Woltersdorf, P. Colombo, and A. Donato, "Structure and Composition of Interlayers in Joints between SiC Bodies," *Journal of the European Ceramic Society*, **17** [10] 1259-65 (1997).

[4]J-H Kim and Y-C Yoo, "Bonding of Alumina to Metals with Ag-Cu-Zr Brazing Alloy," *Journal of Materials Science Letters*, **16** [14] 1212-15 (1997).

[5]K.S. Weil, J.S. Hardy, and J.Y. Kim, in "The Joining of Advanced and Specialty Materials V," Eds.: J.E. Indacochea, J.N. DuPont, T.J. Lienert, W. Tillmann, N. Sobczak, W.F. Gale, and M. Singh, ASM International, Materials Park, OH (2003).

[6]J.Y. Kim, J.S. Hardy, and K.S. Weil, "Effects of CuO content on wetting behavior and mechanical properties of Ag-CuO braze for Ceramic Joinin", *J. Am. Ceram. Soc.*, in review.

[7]A.M. Meier, P.R. Chidambaram, and G.R. Edwards, "A Comparison of the Wettability of Copper-Copper Oxide and Silver-Copper Oxide on Polycrystalline Alumina," *Journal of Materials Science*, **30** [19] 4781-6 (1995).

[8]C.C. Shüler, A. Stuck, N. Beck, H. Keser, and U. Täck, "Direct Silver Bonding - An Alternative for Substrates in Power Semiconductor Packaging," *Journal of Materials Science: Materials in Electronics*, **11** [3] 389-96 (2000).

[9]M.R. Locatelli, A.P. Tomsia, K. Nakashima, B.J. Dalgleish, and A.M. Glaeser, "New Strategies for Joining Ceramics for High-Temperature Applications," *Key Engineering Materials*, **111-112** 157-90 (1995).

COMPUTATIONAL ANALYSIS OF RESIDUAL STRESS FOR Si_3N_4-Al_2O_3 JOINT USING POLYTYPOID FUNCTIONAL GRADIENTS

Caroline S. Lee
Devices and Materials Laboratory
LG Electronics Institute of Technology
Seoul 137-724, Korea

Sung-Hoon Ahn
School of Mechanical and Aerospace Engineering
Seoul National University, Seoul 151-744, Korea

Lutgard C. DeJonghe and Gareth Thomas
Lawrence Berkeley National Laboratory
Berkeley, CA 94720, USA and
Department of Materials Science and Engineering
University of California
Berkeley, CA 94720,USA

ABSTRACT

Alumina and silicon nitride have been joined using a unique approach introducing sialon polytypoids as a Functionally Graded Material (FGM) bonding. The various multilayered FGM samples ranging from three layers to 20 layers were sintered and a crack-free joining of heterogeneous ceramics has been accomplished using the 20 graded layers. Moreover, Electron Probe X-ray Microanalysis (EPMA) scan on a processed crack-free FGM sample shows a smooth concentration profile which verifies interface diffusion during sintering at each graded layer and confirms a successful joining. The thermal stresses of these FGM samples were analyzed using a finite element analysis program (FEAP). These analyses results confirmed how some samples had large residual stresses that resulted in fracture. Therefore, FEAP can be very useful in predicting whether the joints will have cracks or not without fabricating the sample.

INTRODUCTION

The functionally graded materials (FGMs) are promising candidate to obtain properties unavailable in any one homogeneous monolithic material. The design of FGMs is intended to take advantage of certain desirable features of each of the constituent phases. For example, if the FGM is to be used to separate regions of high and low temperature, it may consist of pure ceramic at the hotter end because of the better resistance to the higher temperatures. In contrast, the cooler end may be metallic because of its better mechanical and heat-transfer properties [1]. However, there are factors that reduce the success of such a goal. Stress that arises due to the joining two dissimilar materials is one such factor. As in many joining and composites problems, the effect of residual stress, arising either from processing or from in-service temperature variations, takes on an important role. At a free surface of a joint, the sharp discontinuity in coefficients of thermal expansion (CTE) leads to a signularity, with the linear elastic stresses approaching infinity at the surface [2]. For a material without layers, where the properties vary continuously with position, these sigularities are eliminated [3]. Such a design allows a gradual change in thermal expansion mismatch, minimizing the thermal stresses arising from cooling or heating. Therefore, FGMs offer solution to the thermal stress problem and have created wide interest recently [4-9].

Since one critical design goal and motivation in FGM research is the minimization of thermal stresses, several studies have focused on the theoretical and experimental assessment of these stresses in FGMs. A majority of this analytical work has been for FGM films or other simple structures, for which geometrical assumptions allow for much simplified 1-D linear elastic calculation [1]. For a more general 2-D or 3-D problem, numerical methods such as a finite element analysis (FEA) are required. In this paper, residual stresses are analyzed in the axisymmetric mode, using the computer program FEAP [10]. These formulations do not contain an explicit reference to the microstructure of the FGM, but assume that the behavior of the composite can be calculated solely from the properties and volume fractions of the constituents. By using the effective properties of the FGM and by tracking the changes in the composition through the effective properties, the residual stress calculations are obtainable without specific microstructural modeling. Moreover, since the actual residual-stress measurement and stress analysis of this FGM sample is very difficult, the calculation using FEA will be beneficial to the majority of practical FGM material combinations.

In this paper, various FGM joints were processed and the thermal stresses for each sample were analyzed to confirm experimental results. For the crack-free FGM sample, Electron Probe X-ray Microanalysis (EPMA) was used to scan for the compositional mapping across the length of the sample.

EXPERIMENTAL PROCEDURE

Material Fabrication

The appropriate layers of powder mixes were successively filled in a graphite hot-press die to prepare the sample. The sample was subsequently hot pressed at 50 MPa, at 1700°C, for two hours, and furnace-cooled to room temperature at 2°C/min. Details of the processing and characterization methods are those described by Lee *et al.* [11]. Various numbers of layers and thickness of joints were processed to reduce the residual stress. These joints are discussed in the results and discussion section.

EPMA of Crack-Free FGM Sample

To study how the interface diffusion is taking place at each layer of FGM sample, EPMA was used for the compositional mapping across the length of the sample. Using a full transverse scan, data were taken every 20 μm, with a total of 486 points to make sure that the compositional mapping is done across each interface.

Calculation of Residual Stresses Using the Finite Element Method Program

The thermal stresses of this FGM were computed taking into account of both CTE and modulus variation of the multitude of joining layers. The residual stresses were computed with a finite element method: Finite Element Analysis Program (FEAP) [10]. The details and conditions of the residual stress calculations are described in the paper by Lee *et al.* [12]. The thermal stresses of those various FGM joints were analyzed and discussed in the results and discussion.

Table I. Physical Constants for the materials used for FEAP calculation [12]

Properties	Si_3N_4	Polytypoid	Al_2O_3
E (GPa)	330	290	390
ν (Poisson'sratio)	0.22	0.22	0.22
α ($*10^{-6}/^oC$)	3.6	5.6	8.8

RESULTS AND DISCUSSION

Crack-free Joint

Direct joining of Si_3N_4-polytypoid-Al_2O_3 was fabricated at first to study if residual stress build-up among layers can be minimized by using a thin polytypoid interlayer, which has an intermediate CTE between those of Si_3N_4 and Al_2O_3 [11]. Figures 1 through 4 are optical micrographs showing the cross section of the various joints. Cracks developed due to residual stress building up among the layers [11]. Then, a thicker interlayer which has a similar thickness as the joining materials, was used in the middle as shown in Figure 1 since the thin interlayer did not resolve the problem. This thick interlayer was used possibly to relieve some of the strain energy that formed during the joining of these two materials. However, the result in Figure 1 shows that cracks still developed due to residual stress build up among the layers. For Figure 1, cracks developed in the Al_2O_3 side rather than Si_3N_4 side. As the material cools, Al_2O_3, which has a higher CTE, shrinks in tension mode, whereas Si_3N_4, which has a lower CTE value, shrinks in compression. Since ceramics are weaker in tension than in compression, cracks develop in the Al_2O_3 side perpendicular to the direction of tension. Therefore, the vertical cracks were formed due to a tensile stress generated during cooling cycle in the Al_2O_3 side. However, no horizontal cracks were observed in these samples indicating that there was no delamination issue between the layers. Since adding a thicker layer did not resolve crack problems, grading layers were added between the joining materials to relieve residual stress, as shown in Figure 2. The gradient across the length of the joint was varied by 25 wt% increment in composition but some vertical cracks were still observed, indicating that there is still some residual stress build-up. Large vertical cracks seem to propagate from Al_2O_3 due to the large thermal mismatch between Si_3N_4 and Al_2O_3. Moreover, small cracks were observed in the 100% polytypoid layer toward the Si_3N_4 side. It seems that more gradient layers are needed between 100% polytypoid and 75wt% 12H Polytypoid/25wt% Si_3N_4 as to reduce residual stress. Figure 3 shows the joint, with more added gradient layers between 100% polytypoid layer and Si_3N_4 side to remove little cracks. Most of the small cracks were removed, but large cracks still exist across the FGM, indicating that a finer gradient is needed in the Al_2O_3 side as well. Therefore, from these series of joints that were processed, it is deduced that a more consistent gradient is needed to create a crack-free joint. Moreover, previous studies have indicated that a better result can be obtained by varying the number of layers and the incremental change in composition from layer to layer, than by increasing the overall graded joint thickness [13].

Finally, an FGM joint which consists of 20 graded layers, was introduced by a powder stacking method (Figure 4)[16]. The composition along the gradient was varied by 10 wt% to create a smooth gradient across the thickness. As can be seen from Figure 4, the composite Si_3N_4-polytypoids-Al_2O_3 is crack-free. The thermal residual stress that has been mentioned so far can be calculated using the FEAP, and this analysis will be discussed later in this paper.

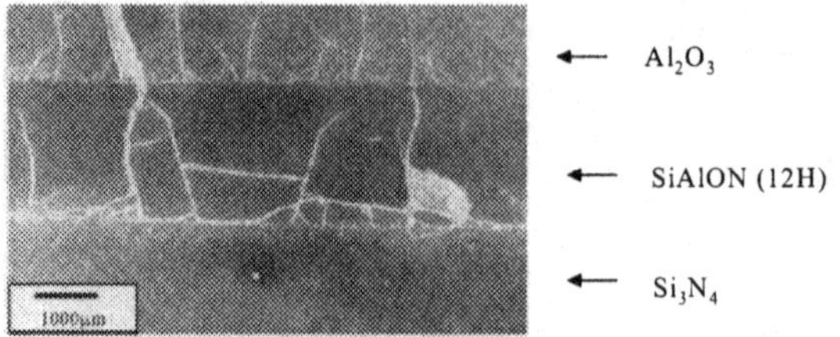

Figure 1. Direct Joining using a thick polytypoid interlayer

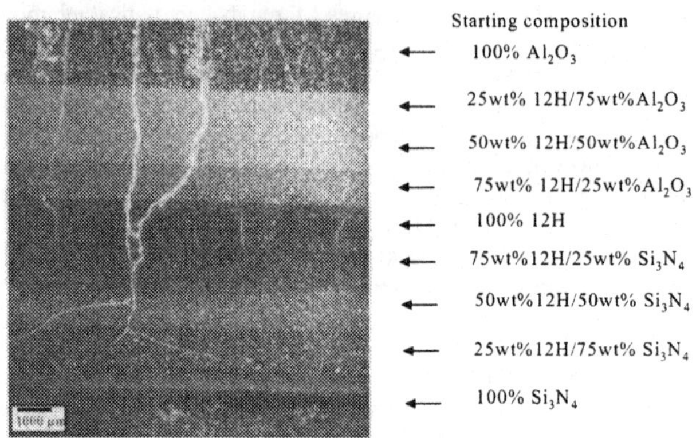

Figure 2. FGM joint between Al_2O_3 and Si_3N_4 with 25 wt% increment in composition

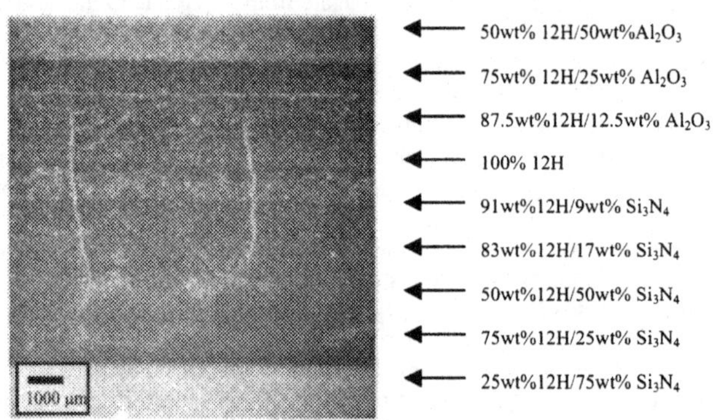

Figure 3. FGM joint between 50wt% 12H/50wt% Al_2O_3 and Si_3N_4 with more gradient layers

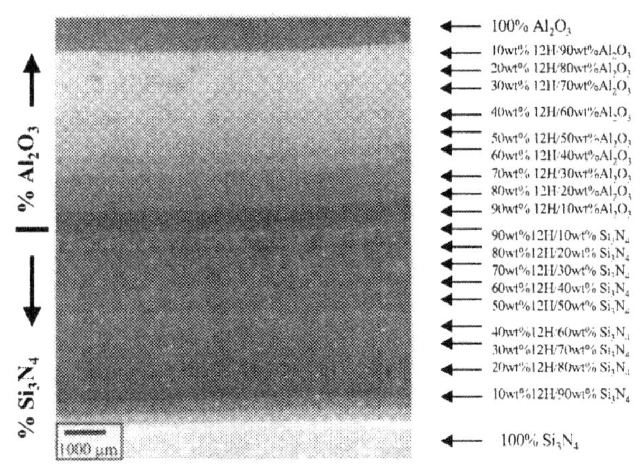

Figure 4. Final FGMs [11]

Interface Diffusion Across the Graded Joint

Compositional mapping using EPMA was done on the crack-free graded joint to study the interface diffusion of each graded layer. Since this FGM was fabricated by the stacking method, each graded layer was visible using optical examination as shown in Figure 4. To verify that interface diffusion did take place during sintering so as to have strong interfaces, EPMA was used to scan the sample across the length of the joint to plot each element as shown in Figure 5. The left side of the graph indicates the Si_3N_4-rich area of the joint, and the right side of the graph indicates the Al_2O_3-rich area of the joint. This figure shows a relatively smooth trend for each element plot, indicating that interface diffusion did take place. If interface diffusion did not occur, the concentration profile would look more stepwise rather than being smooth. Therefore, this EPMA scan was useful in showing a smooth concentration profile to indicate good interface strength.

Residual stress computations

For those various FGM joints that have processed, residual stresses were calculated to correlate with the experimental results. Figures 7-9 show the computed stress distribution of a 3-layer sample with the thick sialon interlayer. As described in previous section, this thicker sialon interlayer was fabricated after using a thin sialon interlayer to join Si_3N_4 and Al_2O_3, showed cracking in the sample. For the axial stress at r=R in the 3 layer joint, the range of stress was found to be from 0.5 MPa/°C to –0.7 MPa/°C. These values are lower than the ones for the 3 layer joint with thin sialon interlayer, but not low enough to result in a crack-free joint, as shown in Figure 1 [12]. Therefore, using thicker sialon interlayers in the joint did reduce some residual stress but not remove cracks completely, as confirmed in the computational analysis. Moreover, it has been found that the grading the layers is important to reduce the residual stress rather than using the same fixed thickness for the single sialon interlayer for abrupt joining between Si_3N_4 and Al_2O_3 [12].

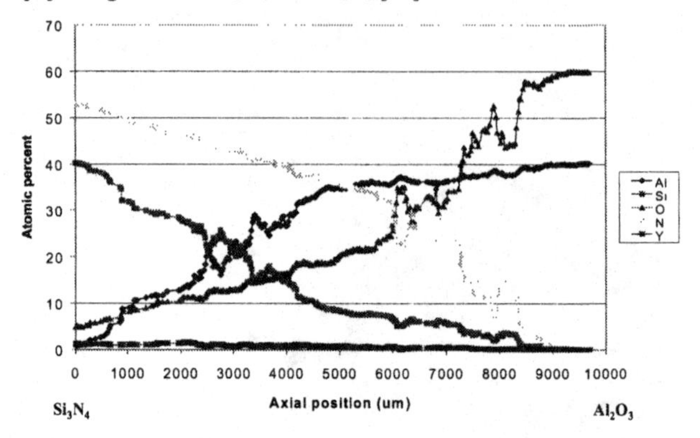

Figure 5. Electron Probe X-ray Microanalysis (EPMA) of crack-free FGM (20 layer) across the length of the sample to study the compositional gradient. Left hand side of the graph shows Si_3N_4-rich area and Al_2O_3-rich area of the joint is shown in the right hand side.

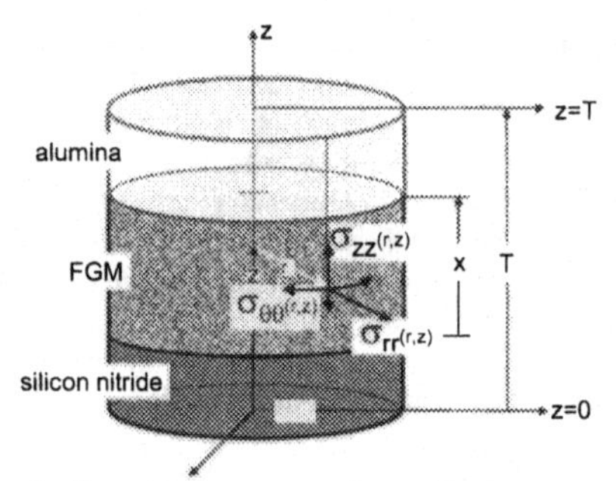

Figure 6. Sample geometry and coordinate systems [12]

After realizing the importance of grading the layer, a series of computational analyses were done to study how the stress is reduced in the graded sample. Figures 10 – 12 show the computed stress distribution of a 5 layer FGM with 50 wt% increment in composition. So the sequence of the starting composition for this graded layer is as follows: 100% Si_3N_4, 50% Si_3N_4/50% sialon (12H), 100% sialon (12H), 50% sialon (12H)/50% Al_2O_3 and finally 100% Al_2O_3. For this sample, the axial stress at r=R in this 5-layer FGM ranged from 0.3 MPa/°C to –0.3 MPa/°C. The stress was reduced significantly compared to that of the 3 layer joint, but more grading of the layer such as 20 layer

FGM reduced the residual stress down to almost minimum [12]. Therefore, this computational analysis is a useful tool to estimate the stresses of various cases without actually processing the sample.

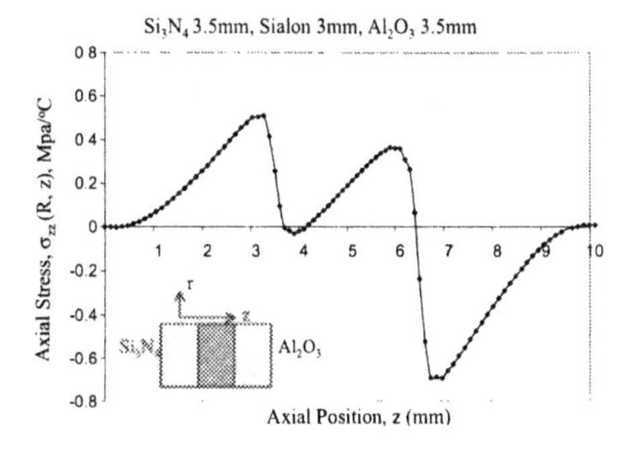

Figure 7. Computed axial stress, σ_{zz} at r=R as a function of z, for a 3-layer sample with thicker sialon interlayer.

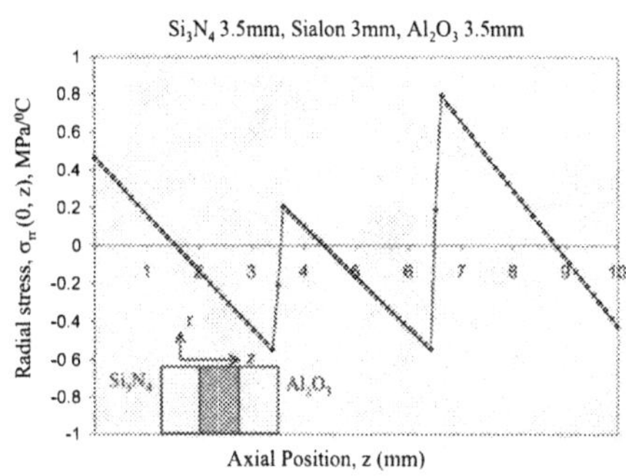

Figure 8. Computed radial stresses, σ_{rr} at r=0 as a function of axial position, z, for the tri-layer sample with thicker sialon interlayer.

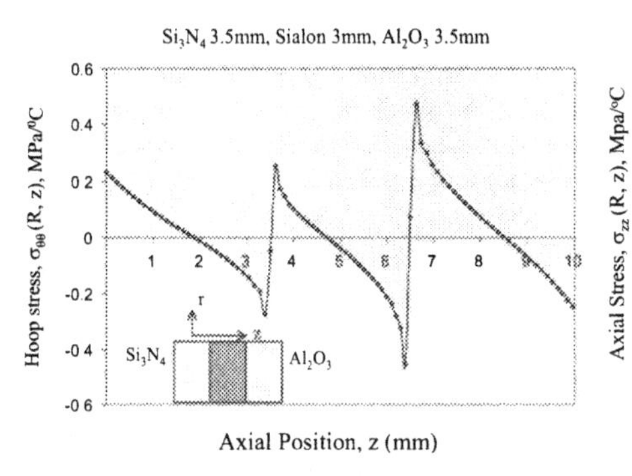

Figure 9. Computed hoop stresses $\sigma_{\theta\theta}$ as a function of axial position, z, at r=R for the trilayer sample with thicker sialon interlayer.

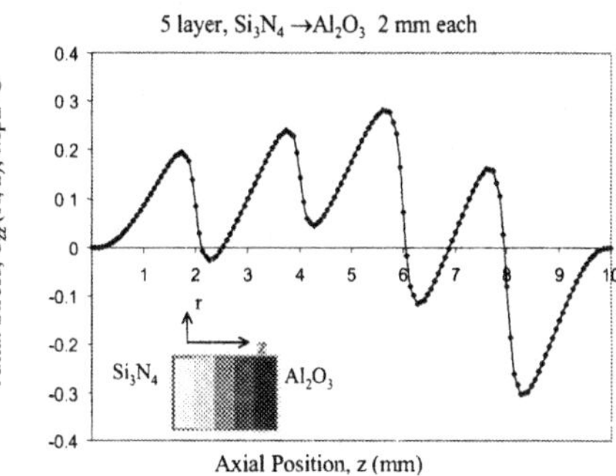

Figure 10. Computed axial stress, σ_{zz} at r=R a a function of z, for the 5-layer sample with sialon interlayer in the middle.

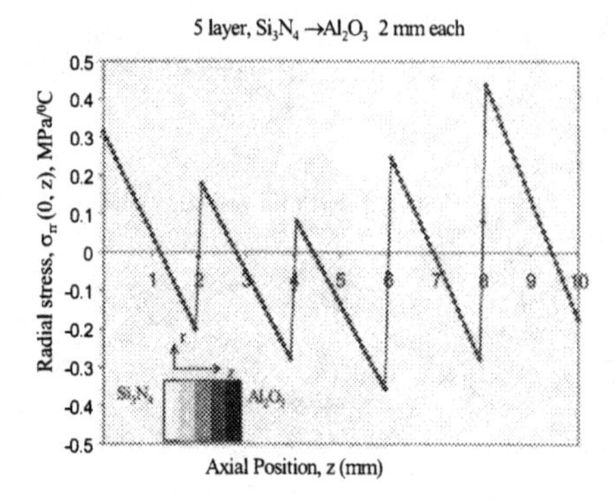

Figure 11. Computed radial stresses, σ_{rr} at r=0 as a function of axial position, z, for the 5-layer sample with sialon interlayer in the middle.

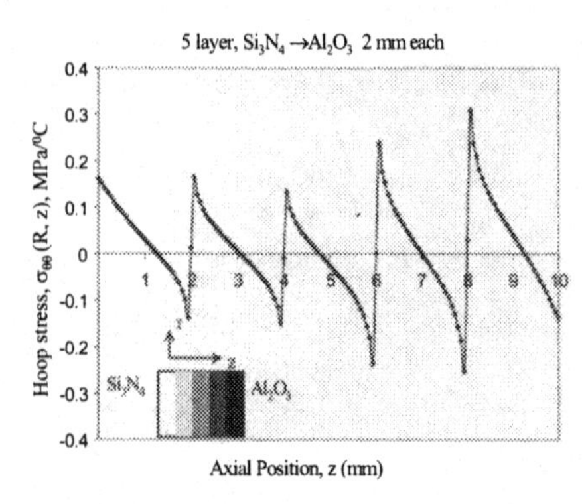

Figure 12. Computed hoop stresses $\sigma_{\theta\theta}$ as a function of axial position, z, at r=R for the 5-layer sample with sialon interlayer in the middle.

CONCLUSIONS

The FGM of Si_3N_4-Al_2O_3 using polytypoid functional gradient was successfully fabricated by powder stacking method through a series of experiments. EPMA scans across the length of 20-layer FGM sample verified that the interface diffusion occurred during sintering at each graded layer by showing a smooth concentration profile. The computational analysis tool (FEAP) was used to calculate the residual stresses in the various FGM samples to compare stress distribution in the samples. The result showed a dramatic decrease in radial, axial and hoop stress as the FGM changes from three layers to 20 graded layers. This analysis explains why a 20 layer FGM was crack-free but other FGMs were cracked. Such analyses are especially useful for graded FGM samples where the residual stresses are very difficult to measure experimentally, and these analyses can also be used in predicting whether these samples have cracks or not before processing the sample.

ACKNOWLEDGEMENTS

This work was supported by the Division of Materials Sciences in the Office of Basic Energy Sciences of the United States Department of Energy. The authors would like to thank Dr. Terrance Becker and Dr. Rowland Cannon at LBNL for help in setting up the FEAP program to calculate residual stress.

REFERENCES
[1]K. Ravichandran, "Thermal Residual stresses in a functionally graded material system", *Mater. Sci. Eng.* **A201**, 269-276 (1995)
[2]D. Munz and Y.Y. Yang, "Stress singularities at the interface in bonded dissimilar materials under mechanical and thermal loading", *J. Appl. Mech.*, **59**, 856-61 (1992)
[3]K. Tanaka, H. Watanabe and Y. Sugano, "A multicriterial material tailoring of a hollow cylinder in functionally gradient materials: scheme to global reduction of thermoelastic stresses", *Comput. Methods. Appl. Mech. Eng.*, **135**, 369-80 (1996)
[4]R. Watanabe, A. Kawakasi and H. Takahashi, "Design, Fabrication and evaluation of Functionally Gradient Material for high temperature use" in Mechanics and Mechanisms of Damage in composites and Multi-Materials, Mechanical Engineering publication, London, 285-299 (1991)
[5]M. Koizumi, "Recent progress of Functionally Gradient Materials in Japan", *Ceramic Engineering and Science Proceedings*, July-August, 333-347 (1992)
[6]B. H. Rabin and R. L. Williamson, "Graded ceramic-metal microcomposites for controlling interface stress" in Microcomposites and Nanophase Materials Eds. D.C. Van Aken, G.S. Was and A.K. Ghosh, TMS-AIME, Warrendale, PA, 103-113 (1991)
[7]R. Watanabe and A. Kawasaki, "Recent development of Functionally Gradient materials for special application to space plane" in Composite Materials Eds. A.T. Di Benedetto, L. Nicolais and R. Watanabe, Elsevier Science, 197-208 (1992)
[8]M. Koizumi and K. Urabe, "Fabrication and application of Functionally Gradient Materials", in Ceramics Today-Tomorrow's Ceramics Eds. P. Vincenzini, Elsevier Science, 1939-1945 (1991)
[9]B. H. Rabin, R. L. Williamson, R. J. Heaps and A.W. Erickson, "Ni-Al2O3 Gradient Materials by powder metallurgy", *Proc. 1st Int. Conf. Advanced Synthesis of Engineered materials*, San Francisco, CA, 175-180 (1992)
[10]O.C. Zienkiewicz and R.L. Taylor, The finite element method, McGraw-Hill, New York, 1987
[11]C. S. Lee, X. Zhang and G. Thomas, "Polytypoid Functional Gradients; Novel joining of dissimilar Ceramics in the Si_3N_4-Al_2O_3 system", *Acta Mater.*, 49, 3775-3780 (2001)
[12]C. S. Lee, L. C. De Jonghe and G. Thomas, "Mechanical properties of polytypoidally joined Si_3N_4-Al_2O_3", *Acta Mater.*, 49, 3767-3773 (2001)
[13]R.W. Messler, M. Jou and T. T.Orling, "A model for designing functionally gradient material joints", *Welding Journal*, **74**[5], S166 (1995)

Joining

JOINING Si_3N_4 TO AN IRON ALUMINIDE ALLOY USING SOFT INTERLAYERS

M. Brochu[1], M.D. Pugh[2] and R.A.L. Drew[1]

[1]Metals and Materials Engineering Dept.
McGill University
3610 University Street
Montreal, Quebec, Canada
H3A 2B2

[2]Mechanical Engineering Dept.
Concordia University
1455 de Maisonneuve blvd,
Montreal, Quebec, Canada,
H3G 1M8

ABSTRACT

During the joining of ceramics to metals, the residual stresses produced during cooling are absorbed through plastic deformation of the metallic component. However, some materials exhibit a limited plastic behavior. Intermetallic materials possess this characteristic. Sound joints between Si_3N_4 and an iron aluminide alloy were obtained with the utilisation of soft interlayer (Cu and Ni). In both cases, a binary Cu-Ti alloy was used as active filler metal. This research focuses on the correlation obtained between the reaction layer development and the mechanical properties. The Cu interlayer was found to dilute the brazing alloy and a better control of the reaction layer growth rate was observed. In the case of the Ni interlayer, changes in the chemical gradients of the diffusion couple dissolve the silicide component of the reaction layer, limiting the possibility of optimization of the mechanical properties.

1. INTRODUCTION

The type of atomic bonding defines the material properties. Ceramic materials are known to possess either ionic, covalent or mixed bonding. In particular, covalent bonding is responsible for the low coefficient of thermal expansion, high hardness observed in nitride ceramics, such as silicon nitride [1]. The free electrons cloud observed in metallic bonding on the other hand, leads to different properties. Therefore, metallic materials have high values of CTE, exhibits higher thermal conductivity and plasticity [1]. Intermetallics are stable compounds possessing a mixture of metallic and covalent or ionic bonding. This mixture gives temperature dependant properties. Iron aluminide (Fe_3Al) possesses a high CTE ($8.2 \times 10^{-6}/^{\circ}C$) but low ductility and high yield stress below the ordering temperature [2].

Industrial applications require silicon nitride/stainless steel joints, where brazed joints are conventionally used. However, the recent developments in non-stoichiometric iron aluminide alloys have demonstrated improved properties (ductility, creep) compared to the pure stochiometric materials [3]. These alloys possesses better high temperature properties than the conventional stainless steel and are slowly replacing the 400 series [4]; thus providing motivation to study this joining system.

Metal-ceramic joints have been studied for more than 20 years. The major problem for the development of reliable joints is the formation of residual stresses induced in the bonded assembly during cooling from the joining temperature as a result of the coefficient of thermal expansion (CTE) and elastic modulus mismatches between the metal and the ceramic [5]. In addition, the strain hardening coefficient, yield stress and interlayer thickness have considerable effects on the residual stress formation and distribution [6]. At certain intensity, these stresses may cause crack initiation in the joint.

The most popular solution to overcome the CTE mismatch between the metal and the ceramic is to insert metallic interlayers having a high level of plasticity in between the parent materials. A combination of a low yield strength and high thermal expansion matching the material to be joined is desired [7]. Copper and nickel interlayers have been found to be more effective than low thermal expansion molybdenum, for the fabrication of strong bonds between Si_3N_4 and steel or nickel-based superalloys [8]. The respective properties of the various element used in different systems is presented in Table 1.

Table 1. Properties of various materials involved in Si_3N_4 to metal joining (annealed condition).

Materials	CTE ($X10^{-6}$ /K)	Young Modulus (GPa)	Yield Stress (MPa)	Ref.
Si_3N_4	3.2	280-300	-	9
Steel (316)	15.8	179	414	10
Superalloy (718)	13	200	414	10
Mo	6	330	400	10
Ni	13.1	207	110	10
Cu	17	115	69	10
FA-129	12.5	140	425	4

The objective of this study was to investigate the microstructure/property relationship developed during partial transient liquid phase bonding of a ceramic-intermetallics system. The materials used in this study were silicon nitride ceramic and FA-129 iron aluminide alloy. The iron aluminide alloys exhibit a low level of plasticity compared to metals, magnifying the residual stress level formed during cooling and therefore requiring the use of soft, ductile interlayers (Cu and Ni).

2. EXPERIMENTAL PROCEDURE

2.1 Starting Materials

Figure 1 presents a sketch of the two sample configurations used.

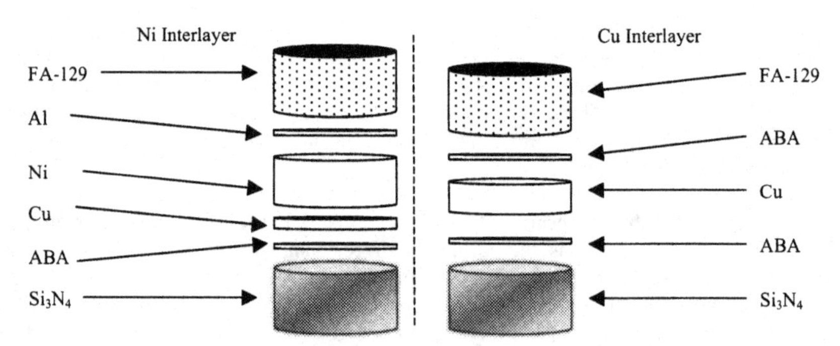

Figure 1. Schematic of sandwich sample design

The silicon nitride ceramic used was is the Ceralloy 147-31N, purchased from Ceradyne. This ceramic contains between 7-8 wt% of sintering additive, mainly yttria and alumina. Ceramic rods of 10 cm length and 8 mm of diameter were cut with a diamond cut-off wheel into 3 mm thick discs. The ceramic was polished to 1μm. The FA-129 alloy (15.9Al, 5.5Cr, 1Nb, 0.05C, bal. Fe) supplied by CANMET, Canada, was received as a sheet of 30cm x 15cm x 0.7cm. Round discs of 3 mm thickness and 8 mm diameter where machined with a diamond tooling. The intermetallic alloy was also polished to 1μm. The active brazing alloy used was a binary composite powder, where the core was composed of titanium and the deposited shell was copper. The built-up shell of the brazing alloy was produce by successive electroless plating deposits performed in a copper sulfate based bath. A more complete description of the procedure is found in [11].

Nickel interlayers of 8mm diameter were punched from a plate of 0.787 mm initial thickness. Both sides where polished to 5 microns. Aluminum discs of 8 mm diameter were punched from commercially available aluminum foil of 30 microns thickness. The copper buffer layer was also punched from foil of 60 μm giving 8 mm diameter discs.

The copper interlayer used in the Cu interlayer system has a thickness of 200 μm and 8 mm diameter. Both side were polished down to 1 μm. A 50μm thick film of brazing paste was deposited between the soft interlayer and the ceramic and the FA-129 alloy. In both systems, all components were cleaned prior to bonding in acetone and ethanol for 15 minutes in an ultrasonic bath.

2.2 Joining procedure

In both cases, the sample was inserted in a graphite jig and put in a graphite-lined furnace. The joining atmosphere used was a slightly positive pressure of argon (99.8% purity). A static load of 300 kPa was put on the sample to maintain alignment and improve the contact at the different interfaces. In both cases, flowing argon was maintained during the cycle.

Double soaking cycle was used in the case of the Ni interlayer. The initial soaking temperature was 950°C for a soaking time of 30 minutes, to allow the formation of a reaction layer at the ceramic interface. The homogenization temperature tested was 1150°C for holding time ranging between 1.5 and 6 hours. The process cycle began by heating at 10°C/minute up to the first soaking temperature (reaction stage). This was followed by a second heating ramp at 5°C/minute up to the homogenization temperature, soaking, and finally, cooling from the diffusion plateau at a rate of 5°C/minute down to 300°C.

A single soak joining cycle was used for the sample with the Cu interlayer. The specimens were heated at a rate of 10°C/min up to the joining temperature and cooled at 5°C/min down to 300°C. The joining temperature studied were 925, 975 and 1025°C for soaking time ranging between 1 and 12 minutes.

2.3 Microscopic evaluation

Cross sections of selected samples were cut with a slow cutting-speed, diamond-wafering saw. The cross sections were mounted and polished down to 0.05 μm colloidal silica. Scanning electron microscopy was performed with a JEOL-840 coupled to an EDAX EDS system. Field

Emission Microscopy on uncoated sample was performed on a Hitachi S4700 coupled with an Oxford EDS detector. The chemical composition was measured with an EPMA JEOL-8900 coupled with a WDS detector. The reaction layer thicknesses were measured by image analysis on a CLEMEX system. A LECO-M-400-2 micro-hardness tester was used to characterize the hardness of the various phases present at the interfaces.

2.4 Strength evaluation

Four-point bending strength was evaluated using a Tinius-Olsen universal testing machine, on cylindrical samples of 5.5 cm length with 8 mm diameter. A similar geometrical arrangement as the microstructure sample was used except that the silicon nitride ceramic was 2.5 cm in length as opposed to 3mm and that the interlayer configuration was reproduced on the other side of the FA-129. The inner and outer spans of the bending jig were 20 and 40mm, respectively. The crosshead speed was 0.5mm/min.

3. RESULTS AND DISCUSSION

The results and discussion section is divided in three main sections: (1) microstructure examination of both interfaces present in the Ni interlayer system, (2) microstructure examination of both interfaces present in the Cu interlayer system and (3) comparison of mechanical properties of both systems.

3.1 Ni Interlayer System: Interfacial Characterization

3.1.1 Si_3N_4/Ni Interface: Figure 2 presents (a) micrograph of the Si_3N_4/Ni interface after the reaction soaking period, (b) micrograph of the Si_3N_4/Ni interface after the homogenisation stage, x-ray line scan of the reaction layer (c) at the early stage of the homogenisation period and (d) the end of the homogenisation period. The results show that after the first soaking stage, the reaction layer is composed of two phases, a Ti and N rich layer adjacent to the Si_3N_4 ceramic and a Ti and Si rich-layer containing a certain level of Cu, between the TiN and Ni interlayer. These layers are commonly observed in reaction zone formed between Si_3N_4 and a Ti-containing active brazing alloy. At this stage of the cycle, no adhesion was observed between the Ni interlayer and the Cu buffer layer (added to prevent rapid diffusion of Ti into the Ni interlayer).

The second holding stage was added to create an interface between with the Ni interlayer and to diffuse the excess of Cu in the interlayer to create a higher melting point interface than prior bonding. However, the second soaking period produces dissolution of the silicide layer formed during the reaction stage. This is observed in the back-scattered micrograph presented in Fig 2(b). In addition, no silicon was detected in the low voltage line scan of the interface on the early stage and after the homogenisation stage, as presented in Fig 2(c) and 2(d). Convolution exists between the Ti-Lα and N-Kα x-ray lines, explaining the behavior of the Ti scan. The Ti and N-rich scans also confirm the absence of Si in the reaction layer, suggesting that the reaction layer is now only composed of TiN. These results suggests that most of the TiN layer growth occurs between the soaking time and the homogenization time (comparison between TiN thickness after the reaction soaking and beginning of homogenisation period). In addition, this rise of temperature increases the solubility limit of silicon into the copper-nickel and dissociation of the Ti_5Si_3 occurs. This releases free Si, which diffuses into the Ni interlayer liberating Ti for reaction with Si_3N_4, leading to further growth of the TiN layer. Within the scattering of the reaction layer growth, the thickness of the TiN layer at the early stage and after the second soaking time were found similar, showing that only the depth of Cu diffusion is changing during this stage.

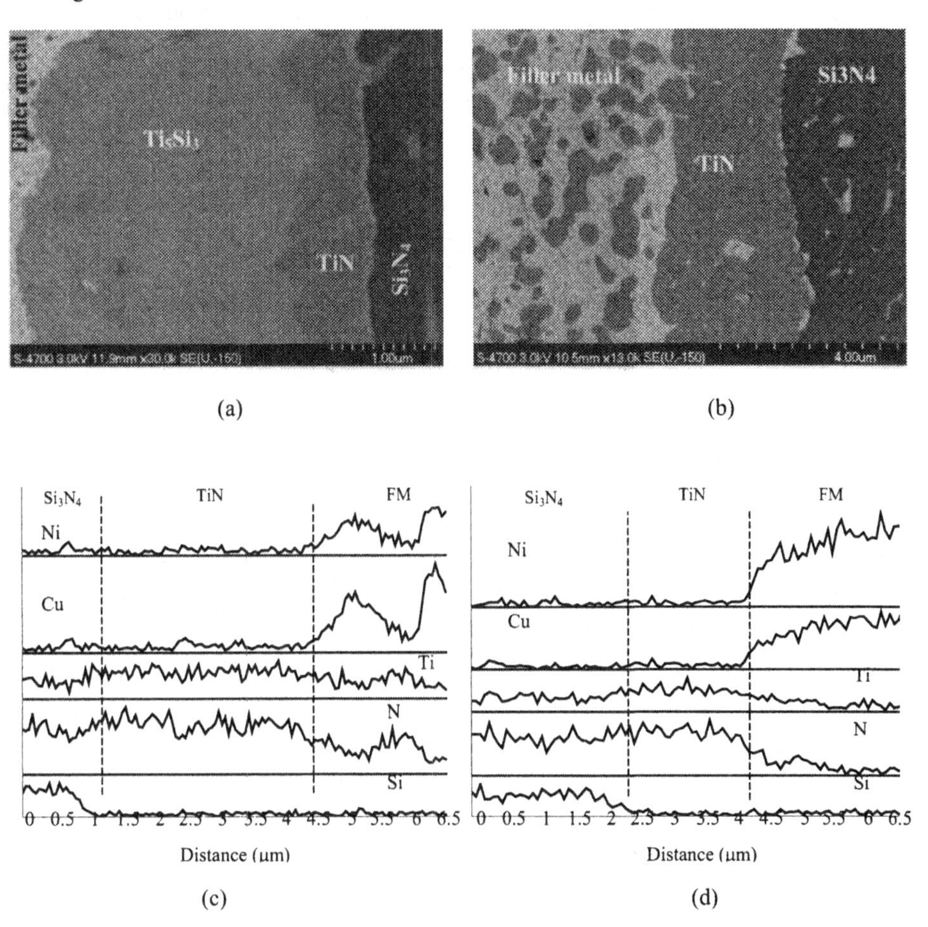

Figure 2. Back-scattered micrographs of (a) Si_3N_4/Ni interface after the reaction soaking period, (b) Si_3N_4/Ni interface after the homogenisation stage, (c) x-ray line scan of the reaction layer at the early stage of the homogenisation period and (d) the end of the homogenisation period.

The second homogenisation period has shown to only influence the diffusion depth of the residual copper. The formation of this solid solution will lead to (1) increase of the yield strength of the interlayer in the vicinity of the ceramic (compared to pure Ni) and (2) diffusion of the low melting point component, Cu, for instance. Obviously, the highest copper concentration was measured near the interface. Table 2 presents the peak copper concentration measured as a function of the homogenization time and the corresponding solidus temperature for the alloy.

The highest concentration was observed for the lower diffusion time. However, even for this lower extreme condition, enough diffusion was achieved and isothermal solidification had occurred. For any sample being joined for longer times or higher temperatures, the copper concentration near the interface is lower, raising the melting temperature of the copper-rich zone of the core interlayer. The re-heating of such a sample even above the joining temperature will not re-form a liquid at the interface, demonstrating that the conditions for PTLPB were achieved. Also, the presence of Cu raised the yield strength of the alloy near the interface to values ranging between 180 and 200 MPa.

Table 2. Chemical concentration of Cu at the vicinity of the reaction layer and corresponding solidus temperature [12].

Time (min)	Wt%Cu	Solidus Temp (°C)
90	50	1255
180	40	1290
270	36	1325
360	32	1330

3.1.1 Ni/FA-129 Interface: Figure 3(a) presents a secondary image of the interface Ni/FA-129 and 3(b) the microprobe line scan of the interface. Identification of the respective material is observed in Fig 3(a). The Ni/FA-129 interface was not bonded for the test interrupted after the reaction stage. However, after the homogenisation stage, a strong interfacial bond was developed. A reaction layer is present and thickening was observed with increase of the holding time during the second soaking period. The EPMA has revealed that in all cases, the reaction layer possesses similar chemical composition, independent of the thickness. The reaction layer contains 71-74wt% of Ni, 18-20wt% Al and 5-9 wt% Fe. By reporting these compositions on the Al-Cu-Fe ternary phase diagram [12], the phases of the reaction layer were identified to be Ni_3Al-NiAl.

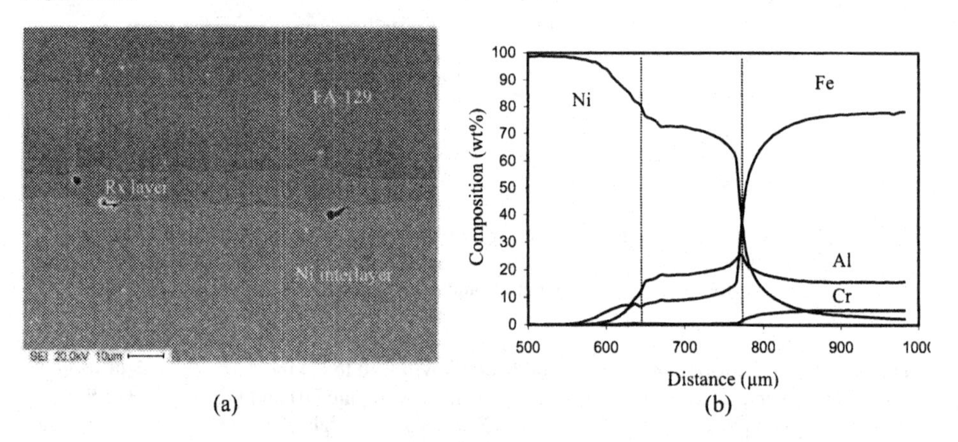

(a) (b)

Figure 3. (a) Micrograph of the FA-129/Ni interface and (b) the corresponding EPMA line scan of the interface.

Vickers microhardness measurements of the reaction layer were taken and results of 374±22 HV were obtained. Tan et al. have determined that the hardness of a Ni-Fe-Al ternary β phase is ≈380 HV for an intermetallic with a similar composition to the reaction layer [13]. This result reinforced the identification of the phase present in the reaction layer.

The formation of the intermetallic phase at the Ni/FA-129 interface is believed to occur through the diffusion of Al from the FA-129 due to the compositional gradient. Mehrer et al. have demonstrated that Al diffuses faster than Fe in the ordered state (atomic structure present at the homogenization temperature [14]) and that the same behavior was found in Ni_3Al intermetallics [15]. In addition, the residual thickness of the Ni interlayer measured through EPMA line scan is relatively constant compared to its initial values, suggesting that the dissolution of the iron aluminide alloy combined with the diffusion of Al from the FA-129 to the Ni interlayer is the preferential mechanism for the diffusion layer formation. The presence of intermetallics at the grain boundaries of the iron aluminide (arrows in Fig 3(a)) also reinforces the statement that the FA-129 dissolved along the grain boundaries.

3.2 Cu Interlayer System: Interfacial Characterization

3.2.1 Si_3N_4/Cu Interface: Figure 4(a) presents a back-scattered micrograph of the interface between the Si_3N_4 and the Cu interlayer for a sample brazed at 1025°C for 1 minutes and (b) the corresponding x-ray line scan of the interface (dashed line in Fig 4(a)). Independently of the joining conditions, the reaction zone is composed of two layers but the heat treatment parameters affect their relative thicknesses. The low voltage line scan indicates a layer rich in Ti and N adjacent to the interface of the ceramic and a layer rich in Ti and Si and containing Cu next to this layer. The Ti and N-rich layer is believed to be TiN and the Ti-Si is believed to be a Ti_5Si_3 compound containing a certain level of Cu in solution. Silicon has been detected in the filler metal and its concentration decreases further from the reaction layer. Again, the similarity between the Ti and N line scan is produced by the convolution of the characteristic x-rays peaks (Ti Lα and NKα).

The effect of the temperature on the morphology of the layers has shown that both layers grow and the thickness ratio between the TiN layer and the total reaction layer thickness slightly increases with increasing temperature. As the reaction layer growth is a more complex phenomenon than a single reaction system - due to evidence of growth competition between the different reactions - the total reaction layer thickness was measured as a function of brazing parameters and the results are presented in Figure 5. In all three tested temperatures, the data fit the final part of a parabolic trend, suggesting that the growth follows a diffusion-controlled type of reaction. In addition, the fit suggests that in all cases, the growth rate is diminishing to zero, indicating that the reaction layer is not growing fast with respect to time.

3.2.2 Cu/FA-129 Interface: Figure 6 presents a back-scattered electron image of the FA-129/Cu interface from a joint bonded at 1025°C for 12 minutes. This joining condition is the most severe in terms of joining temperature and soaking time. The intermetallic is observed on the left hand side of the micrograph. The dissolution of the intermetallic grain as well as the formation of different precipitates can easily be observed. Copper has penetrated into the intermetallic grains and is observed as copper-rich regions or larger islands. These copper zones are depleted in iron. A very small amount of titanium, from the active brazing alloy has also penetrated into the grains. The titanium penetration is complementary to the iron content and corresponds to the depletion in aluminum. The two different types of precipitates observed are phases rich in iron and titanium (with different stoichiometry) with traces of aluminum embedded in a copper-aluminum matrix.

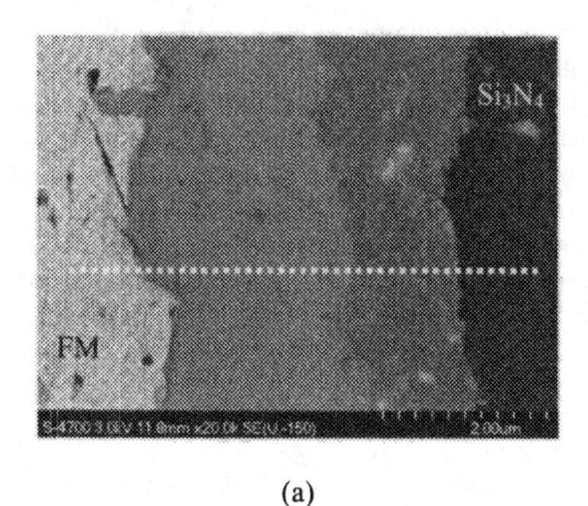

(a)

FM
Ti_5Si_3 + Cu
TiN
Si_3N_4
Cu
Ti
N
Si
Counts
0 0.5 1 1.5 2 2.5 3 3.5 4 4.5 5 5.5 6 6.5 7 7.5
Distance (μm)

(b)

Figure 4. Back-scattered micrographs of the interface between the Si_3N_4 and the Cu interlayer for sample brazed at (a) 1025°C for 1 minutes and (b) low voltage line scan of the reaction layer observed in (a).

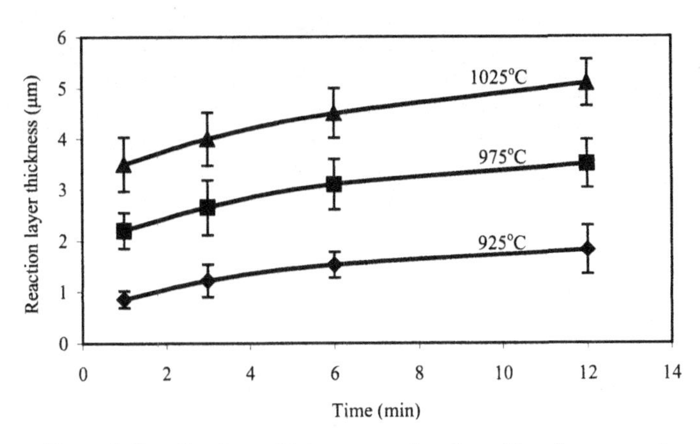

Figure 5. Reaction layer thickness as a function of brazing parameters.

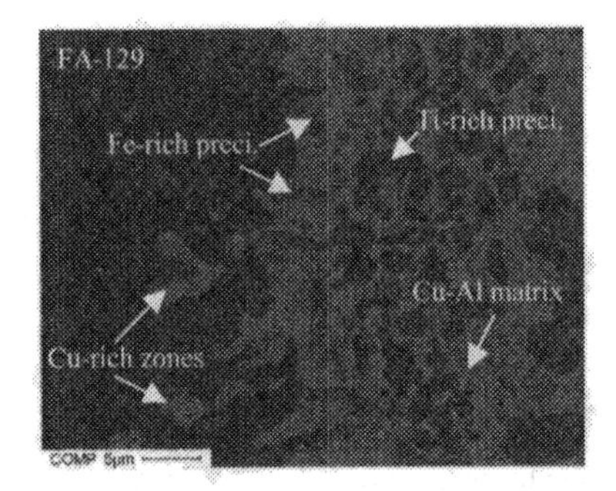

Figure 6. Back-scattered image of the interface FA-129/Cu

Figure 7 presents a back-scattered montage of an FA-129/Cu interface joined at 975°C for 3 minutes. No major phase difference in the precipitates can be observed near to or far from the interface. EDS spectrum performed at different locations (A, B and C) to evaluate the change in composition are presented in Fig 7(b) to 7(d). The elemental composition of the precipitates close to the interface is presented in Fig 7(b) and for a similar precipitate but at a distance of approx. 100 microns from the interface, is presented in Fig 7(c). The height of the energy peak is similar and the comparative ratio of the heights between the different elemental peaks are similar which indicates a very similar chemical composition. Aluminum was still detected (Fig 7(d)) at nearly 100 microns from the interface.

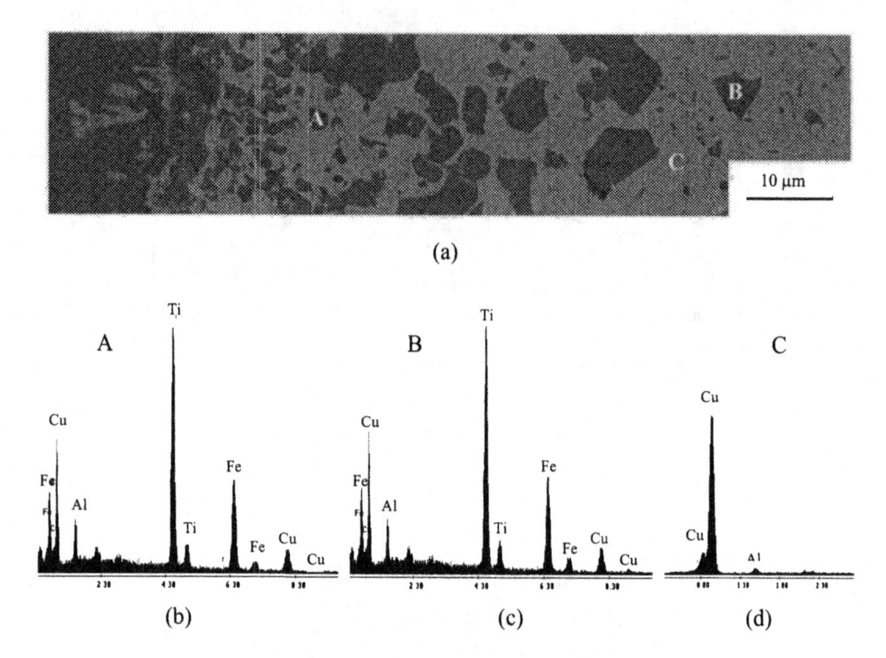

Figure 7. (a) Back-scattered montage of the FA-129/Cu interlayer interface, (b) and (c) EDS analysis of Fe-Ti precipitates and (d) EDS analysis of Cu-Al matrix, respectively

3.3 Comparison of Joint Mechanical Properties

Figure 8 presents a comparison of the optimized four point bending results obtained for the Ni and Cu interlayer system. The bend testing results for brazed Si_3N_4 to Si_3N_4 joint with the same active filler metal was added for comparison purposes. The mechanical property of a ceramic-metal joint is influenced by the thickness of the various reaction layers developed during the joining cycle and the level of residual stresses generated during cooling. In both soft interlayer systems, the failure occurred at the interface. As presented earlier, a partial dissolution of the reaction layer occurs in the Si_3N_4-Ni interface and the final reaction layer thickness was found to be independent of the homogenisation period. In all cases, a bending strength of 80 MPa was obtained.

The optimized bending strength for the Cu interlayer system was 160 MPa, which is the double the strength of the Ni interlayer system. In that case, a conventional reaction layer evolution was observed. For shorter bonding time or lower bonding temperature, the strengths obtained were lower. The reaction layer thickening have shown that increase in soaking time will not lead to major thickening of the reaction layer. Therefore, time is an ineffective bonding parameter in this joining system. The temperature at which the optimised strength was obtained was found to be the highest operational temperature possible as deterioration of the Cu/FA-129 interface occurs if the temperature is raised.

The self-brazed Si_3N_4 results are included to demonstrate that by minimizing the effect of the residual stresses formed during cooling (similar material brazing), respectable strength values were obtained for the active brazing alloy used. In this case, failure occurred through the ceramic, showing that strong interfacial bonding was achieved.

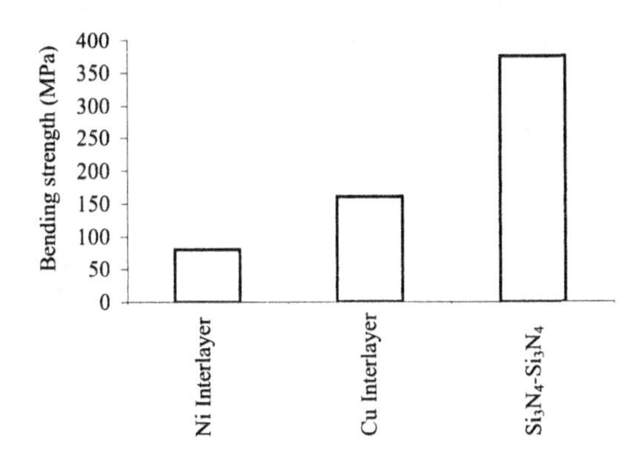

Figure 8. Comparison of four-point bending strengths obtained for Ni and Cu interlayers (self-brazed Si_3N_4 is included for comparison).

4. CONCLUSIONS

This paper has demontrated the feasibility of bonding silicon nitride ceramic to an iron aluminide alloy using partial transient liquid phase joining. The results have shown that similar joint microstructure is developed during bonding at the ceramic interface. The presence of a soft interlayer was effective to reduce the residual stresses developed during cooling. However, the low mechanical properties obtained when compared to ceramic-metal systems demonstrates that further work is required to completely eliminate the detrimental effect of residual stresses created during cooling.

5. ACKNOWLEDGEMENTS

The authors would like to acknowledge Dr. Vinod K. Sikka and Dr. Benoit Voyselle for supplying the FA-129 alloy. The authors would like to thank NSERC for overall funding of the project and FQRNT and McGill University for various scholarships.

6. REFERENCES

1. M. Schwartz, Handbook of Structural Ceramics, McGraw-Hill, New York, 1992.
2. Structural Applications of Intermetallics Compounds, edited by J.H. Westbrook and R.L. Fleischer, Wiley, New York, 2000, 317 p.
3. V. K.Sikka, " Development of Nickel and Iron Aluminides and Their Applications" *Advances in High Temperature Structural Materials and Protective Coatings* pp.282-295 Edited by A.K. Koul et al., National Research Council of Canada, Ottawa, 1994.

4. V.K. Sikka and C.T. Lui, "Iron Aluminide Alloys for Structural Use", *Materials Technology*, **9** 159-162 (1994)
5. S.D. Peteves and M.G. Nicholas, "Materials Factors Affecting Joining of Silicon Nitride Ceramics", *Metal-Ceramic Joining*; pp.43-65, Edited by P. Kumar and V.A. Greenhut The Minerals, Metals & Materials Society, Warrendale, 1991.
6. M. Gao and F. Bao, "Brazing of Silicon Nitride and Steel with Cu-based filler metals: Mechanism and Residual Stress in its joints", *Materials Science Forum*, **189-190** 399-404 (1995).
7. B.T.J. Stoop and G.D. Ouden, "Diffusion Bonding of Silicon Nitride to Austenitic Stainless Steel with Metallic Interlayers", *Metallurgical and Materials Transactions A*, **26A** 203-208 (1995).
8. S.D. Peteves, G. Ceccone, M. Paulasto, V. Stamos and P. Yvon, "Joining Silicon nitride to Itself and to Metals", *Journal of Materials*, **January**, 48-52 74-77 (1996).
9. R.W. Cahn, P. Haase and E.J. Kramer (editors), *Structure and Properties of Ceramics, Materials Science and Technology*, Volume 11, VCH Publishers, 1994.
10. www.matweb.com
11. M. Brochu, M.D. Pugh and R.A.L. Drew, "Active Brazing Alloy Produced by Electroless Plating Technique", *Ceramic Engineering and Science Proceedings*, **23** [3] 801-808 (2002).
12. ASM Handbook Vol 3, Phases Diagrams (ASM International, USA), CD-ROM.
13. Y. Tan, T. Shinoda, Y. Mishima and T. Suzuki, "Stoichiometry Splitting in the Ni-Fe-Al Ternary β Phase", *Journal Japan Institute of Metals*, **57** [7] 840-847 (1993).
14. O. Ikeda, I. Ohnuma, R. Kainuma and K. Ishida, "Phase Equilibria and Stability of Ordered BCC Phases in the Fe-rich Portion of the Fe-Al system", *Intermetallics*, **9** 755-761 (2001).
15. H. Mehrer, M. Eggersmann, A. Gude, M. Salamon and B. Sepiol, "Diffusion in Intermetallic Phases of the Fe-Al and Fe-Si systems", *Materials Science and Engineering*, **A239-240** 889-898 (1997).

GLASS SEALING IN PLANAR SOFC STACKS AND CHEMICAL STABILITY OF SEAL INTERFACES

Z. Yang*, G. Xia, K.D. Meinhardt, K. S. Weil, and J.W. Stevenson
Pacific Northwest National Laboratory, Richland, WA 99352

ABSTRACT

In intermediate temperature planar SOFC stacks, the interconnect, which is typically made from cost-effective oxidation resistant high temperature alloys, is typically sealed to the ceramic PEN (Positive electrode-Electrolyte-Negative electrode) by a sealing glass. To maintain the structural stability and minimize the degradation of stack performance, the sealing glass has to be chemically compatible with the PEN and alloy interconnects. In the present study, the chemical compatibility of a barium-calcium-aluminosilicate (BCAS) based glass-ceramic (specifically developed as a sealant in SOFC stacks) with a number of selected oxidation resistant high temperature alloys, as well as the YSZ electrolyte, was evaluated. This paper reports the results of that study, with a particular focus on Crofer22 APU, a new ferritic stainless steel that was developed specifically for SOFC interconnect applications.

INTRODUCTION

Solid oxide fuel cells (SOFCs), which convert chemical energy of a fuel to electricity via an electrochemical reaction, have become increasingly attractive to the utility and automotive industries for a number of reasons, including their high efficiency and low emissions. Among the different SOFCs, the planar type, as exemplified in Figure 1, is expected to be a mechanically robust, high power-density and cost-effective design. Functionally, the interconnect in a planar SOFC stack acts as a separator plate, physically separating the fuel at the anode side from the air at the cathode side. Thus, to allow the SOFC stack to function well, the interconnect has to be sealed to the adjacent components, such as PENs (Positive cathode-Electrolyte-Negative anode) or metallic frames which hold the PENs. The seals not only prevent mixing of fuel and air, but also keep the fuel from leaking out of the stack. Thus, seal performance not only greatly affects the structural integrity and stability of the stack, but could also dictate the overall stack performance.

With the steady reduction in SOFCs operating temperatures to the intermediate range of 650~800°C [1,2], oxidation resistant high temperature alloys have become favorable materials for interconnect components due to their low raw material and manufacturing cost [3-8]. Among these alloys, some cost effective ferritic stainless steels offer good thermal expansion matching to the ceramic PEN in the stacks. Among the ferritic compositions, chromia forming alloys appear more promising than alumina formers mainly due to the higher conductivity of the oxide scale formed on the surface of the alloy during SOFC operation [9-10]. The sealing between stack components is often carried out using a glass-ceramic (i.e., bonding is achieved by heating a glass to a suitable softening point; the glass then devitrifies during SOFC stack operation to form a mixture of crystalline phases), though other sealing technologies are also under consideration by different developers. In order to maintain the structural stability and minimize the degradation of SOFC performance, the sealing materials have to be chemically compatible with the alloy interconnect.

* zgary.yang@pnl.gov; Tel: (509) 375 3756.

In this study, a barium-calcium-aluminosilicate (BCAS) based glass specifically developed as a sealant in SOFC stacks [11] was chosen as an example. Its chemical compatibility with selected alloy candidates from different groups of traditional oxidation resistant high temperature alloys has been systematically evaluated. Results for several traditional alloys were published previously and the mechanisms regarding the interactions between the sealing glass and alloy interconnects were interpreted as well [12,13]. This paper discusses the chemical stability of BCAS glass-ceramic seal interfaces with YSZ electrolyte and chromia-forming ferritic stainless steels, with particular emphasis on Crofer22 APU, a new ferritic stainless steel that was developed specifically for SOFC interconnect application [14]. For purposes of comparison, the chemical compatibility of the sealing glass with traditional chromia-forming ferritic stainless steels will also be discussed.

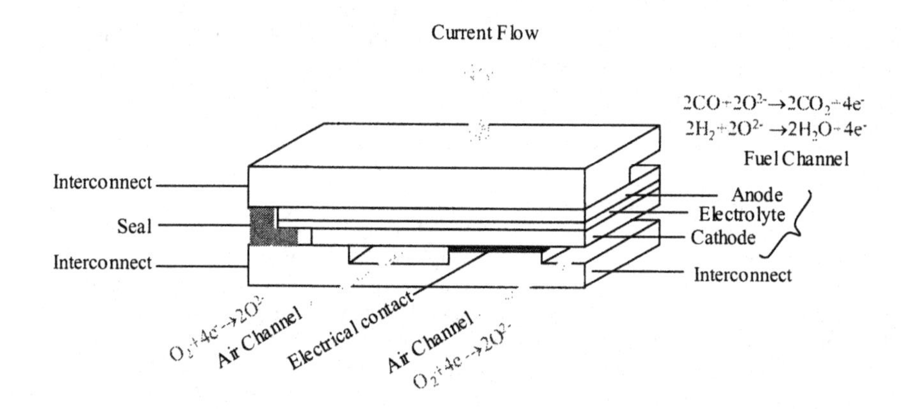

EXPERIMENTAL

The candidate alloys involved in this study include AL29-4C, 446 and Crofer22 APU. Their chemical compositions are listed in Table I. Alloy coupons with a thickness of 0.5~1.0 mm was first cut into 12.7mm x 12.7mm squares, and then ground and polished using 600 grit SiC paper. Before joining, the polished alloy coupons were ultrasonicated in alcohol for 10 minutes, and then rinsed with acetone. The electrolyte coupons were made by aqueous tape casting of 8 mol% yttrium stabilized zirconia (YSZ) obtained from Zirconia Sales (America). The tape was formed by mixing an aqueous slurry of zirconia powder with an acrylic emulsion binder (Rohm&Haas) and cast on a Mylar carrier film to dry. The dry tape was cut into coupons and sintered at 1350°C, cooled, and then creep flattened at 1350°C.

Table I. Alloy compositions

Alloys	Cr	Fe	C	Mn	Si	Al	Ti	P	S	Zr	Re
446	25	Bal.	0.20	1.50	1.00	--		0.04	0.03	--	--
Fecralloy	22	Bal.				5.0	--			0.10	0.10 Y
Crofer 22 APU	22.8	Bal	0.005	0.45	--	--	0.08	0.016	0.002	--	0.06 La
AL29-4C	29.0	Bal	0.01	0.3	0.2	--	--	0.025	0.02	--	--

* Crofer and AL are trademarks of ThyssenKrupp and Allegany Ludlum, respectively.

The sealing glass, designated as G18, is a BCAS based-glass that has the following composition as analyzed by inductively coupled plasma (ICP): BaO (56.10%), CaO (7.19%), Al_2O_3 (5.39%), B_2O_3 (6.66%), and SiO_2 (21.45%). This glass was specifically developed at PNNL for SOFC stack sealing applications [12]. In addition to the designated oxide components, analysis also revealed impurities of K_2O (0.26%), Li_2O (0.06%), NiO (0.04%), MgO (0.36%), Na_2O (0.43%), SO_3 (0.25%), SrO (0.11%), Fe_2O_3 (0.05%), Y_2O_3 (0.08%), ZrO_2 (1.56%), and dissolved water. The glass was produced by melting the raw material oxides at 1400°C for 60 minutes in a platinum crucible, followed by roller quenching after a temperature reduction to 1250°C. Oxygen was injected to stir the melt for the duration of the melt time. The glass frit was subsequently dry-milled in an alumina lined jar with alumina media to <15 μm particle size. The resulting powder was then attrition-milled with zirconia media in isopropanol to an average particle size of 1 μm and dried before tape casting. The 0.8 mm thick tapes were formed from organic-based tape-cast 0.2 mm sheets which were laminated to the final thickness.

To prepare a joined "couple," the glass tape was sandwiched between two identical alloy coupons or an alloy coupon and YSZ plate, as shown in Figure 2 (a), and then placed into a loading fixture in an electric furnace. A dead load of 6.9×10^3 Pa was applied on top of the joints. The joining was carried out by heating in air to 850°C for 1 hour, followed by a dwell at 750°C for a designated period of time before cooling. During the heat treatment the glass softened and bonded to the alloy coupons and YSZ electrolyte. As some devitrification (formation of crystalline species) also occurred during the heat treatment, the G18 will be referred to as a "glass-ceramic" material. The heating and cooling were controlled at 2°C/min and 1°C/min, respectively. The couples were subsequently epoxy-mounted and sectioned. Polished cross-sections were examined via scanning electron microscopy (JEOL model 5900LV) and energy-dispersive X-ray analysis (EDX) at an operating voltage of 20 kV.

RESULTS and DISCUSSION

Interfaces in the Alloy-Glass-YSZ Seal

Through heat treatment in air (850°C for one hour followed by 750°C for 4 hours), a coupon of AL29-4C was joined to an identically sized YSZ plate by the sealant G18 to form a joined couple, as shown schematically in Figure 2 (a). The joined couple was then cross-sectioned at the middle and analyzed by SEM (refer to Figure 2(b)). A scan of the entire cross-section revealed an inhomogeneous microstructure along the AL29-4C/G18 interface, but a homogeneous one along the G18/YSZ interface. At the edges of the joined couple, it appeared

Table II. Results of X-ray energy dispersive analysis at points marked in Figure 2

Elements (atomic%)	O	Al	Si	Ca	Ba	Zr	Y
Point 1	65.99	--	--	--	--	29.29	4.72
Point 2	69.56	0.73	4.24	1.30	4.78	19.39	--

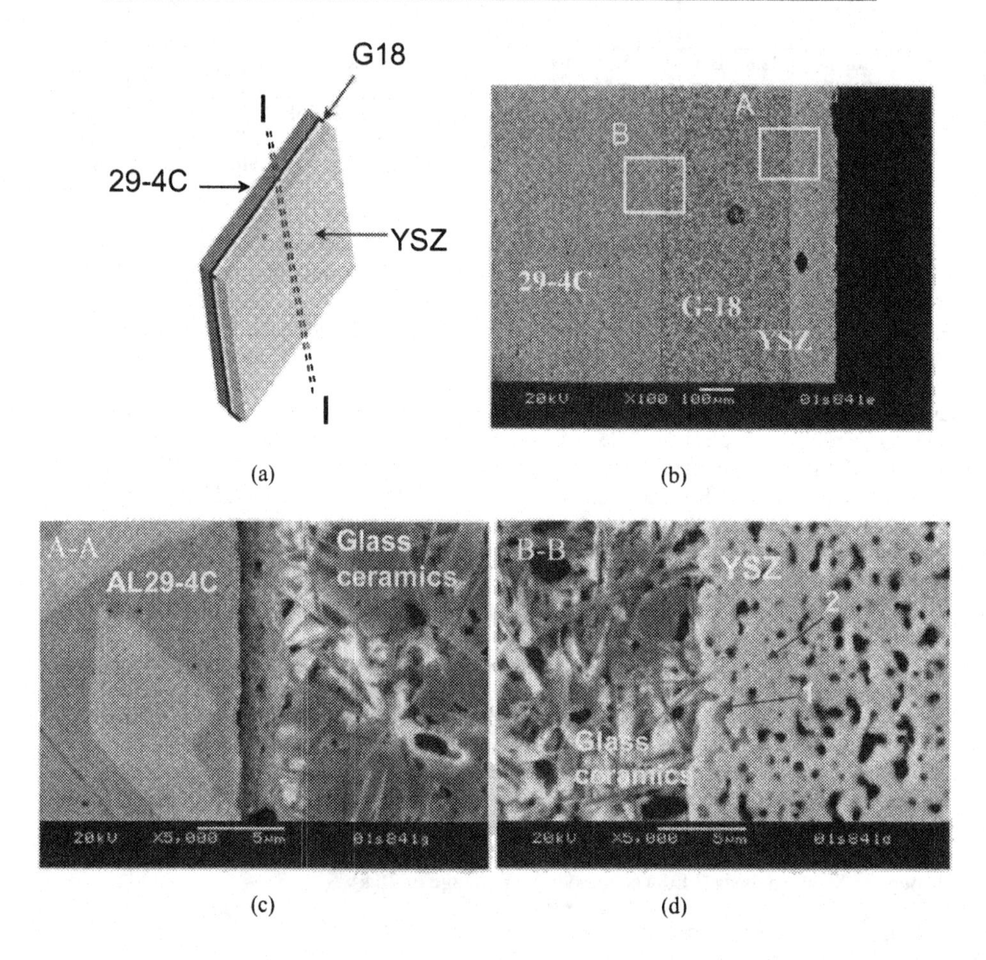

(a) (b)

(c) (d)

Figure 2. (a) A schematic of a 29-4C/G18/YSZ couple, and SEM images of the interfacial cross-section at: (b) in the interior area, (c) from the region marked as "B", and (d) the region marked as "A" in (b). The 29-4C coupon (12.7mm x 12.7mm x 0.5 mm) was joined to the YSZ plate (12.7mm x 12.7mm x 0.5mm) with G-18 via heat treatment at 850°C for one hour, followed by 750°C for four hours in air.

that the AL29-4C reacted with G18 during heating to form a yellowish product ($BaCrO_4$, as discussed in the following section) along the AL29-4C/G18 interface, leading to separation of the sealing glass-ceramic and the stainless steel. In the interior area, as shown in Figure 2 (c), the sealing glass-ceramic appeared to be well-bonded to the stainless steel, with a clearly discernible steel/glass-ceramic interface. At this interface, the glass-ceramic interacted with the stainless steel, forming a reaction zone (see discussion below). The glass-ceramic was also well-bonded to the YSZ along the G18/YSZ interface, as shown in Figure 2 (d), with no significant reaction between the 2 materials. Some penetration of the glass-ceramic into the YSZ was evident, which likely resulted in strong bonding between the glass-ceramic and the YSZ plate via mechanical interlocking. EDS point analysis (refer to Figure 2 (d) and Table II) was used to estimate the glass composition in the pores near the YSZ surface.

Given the fact that the glass-ceramic/alloy interface appeared to be weaker and more complicated than the glass-ceramic/YSZ interface, due to the chemical interactions occurring at that interface (and, in some cases, the relatively weak oxide scale adherence to the alloy substrates), the balance of this study focused on glass-ceramic/alloy interfaces.

Chemical Stability of the Interface between G18 and Traditional Chromia Forming Alloys

The ferritic stainless steel 446 was selected as an example of traditional chromia forming alloys; its chemical stability with the glass-ceramic is briefly discussed here. As schematically shown in Figure 3 (a), two identical 446 coupons were joined by the sealant G18 to form a joined couple via heat treatment in air at 850°C for one hour, followed by 750°C for 4 hours. The joined couple was then cross-sectioned at the middle and the interface of the joint was analyzed on SEM. A scan of the entire cross-section revealed an inhomogeneous microstructure along the glass/446 interfaces. Accordingly, detailed SEM analysis was carried out on different locations from the edge to the inside area on the interfaces. Figure 3 (b) shows a typical secondary electron image taken from the edge area of the joint couples. An enlarged image from the area marked as "C" is further shown in Figure 3 (d). The images from the edge area clearly indicate that the sealant G18 reacted with the ferritic stainless steel 446 to form a yellowish compound, creating gaps between the glass-ceramic and stainless steel coupons after heat treatment. Detailed structural and chemical analyses [12] confirmed that the formed compound was barium chromate ($BaCrO_4$) that was formed from a reaction between the barium oxide in G18 and the chromia scale grown on chroming forming alloys during heating via the following reaction:

$$2Cr_2O_3(s) + 4BaO(s) + 3O_2(g) = 4BaCrO_4(s) \quad (1)$$

The fact that this reaction requires the presence of oxygen or air explains why the $BaCrO_4$ was only found at the edge or near edge areas of the join, where oxygen in the air was easily accessible. Due to the large thermal expansion mismatch between barium chromate and G18 or 446[12,18], the extensive formation of barium chromate probably resulted in the observed gap between the sealing glass and alloy coupons.

In additional to the above solid-state barium chromate formation reaction, the dominant chromia vapor species when moisture is present, chromium oxyhydroxide ($CrO_2(OH)_2$) [15-17], can also react with barium oxide in the sealing glass to form barium chromate via the following reaction:

$$CrO_2(OH)_2(g) + BaO(s) = BaCrO_4(s) + H_2O(g) \quad (2)$$

This reaction mechanism helps explain the observed homogenous barium chromate formation on the surface of the glass-ceramic that was squeezed out during the joining heat treatment.

In the interior of 446/G18/446 joint, as shown in Figure 3(c), there was no $BaCrO_4$ formation, and the sealing glass was bonded with the ferritic stainless steel. A chromium rich layer was observed in the glass-ceramic at the interface, presumably due to dissolution of chromium from the alloy scale into the glass-ceramic to form chromium-rich phases. The G18/446 interface in the interior area was also characterized by the formation of porosity along the interface, as shown in Figure 2 (c). These pores were likely created through the formation of vapor species via the interaction of alloy elements, especially chromium, with dissolved water and alkaline oxide residues in the glass [13].

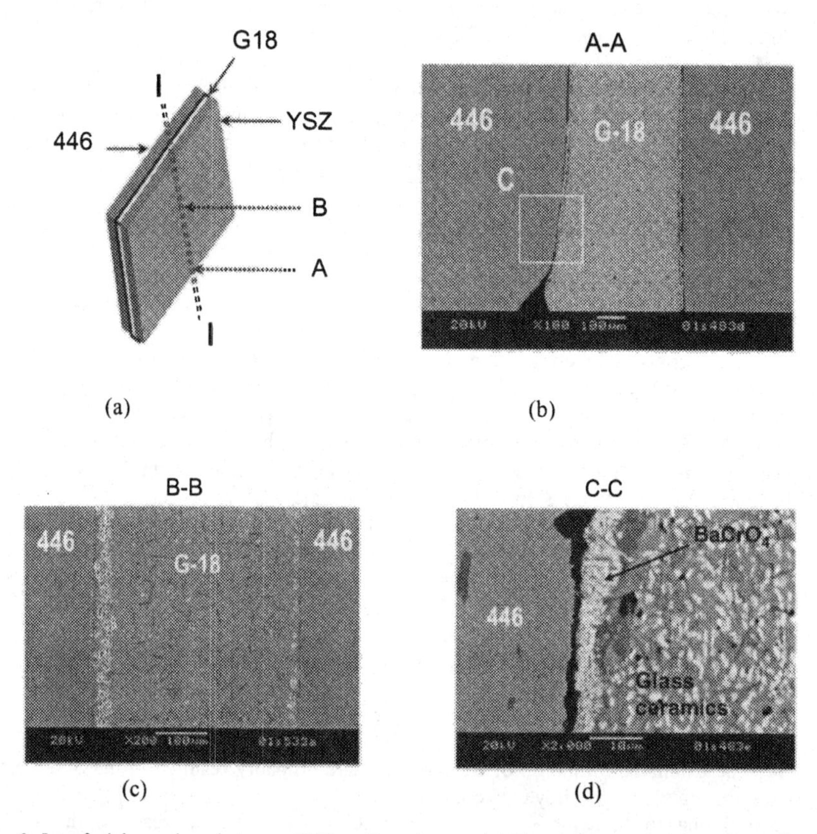

Figure 3. Interfacial reactions between G18 sealing glass and 446 stainless steel: (a) a schematic of the joined couple (446/G18/446), and SEM images of the interfacial cross-section at: (b) the edge area A, (c) the interior region, and (d) from the region marked as "C" in (b). The 446 coupons (12.7mm x 12.7mm x 0.5mm) were joined to the G18 through heat treatment at 850°C for one hour, followed by 750°C for four hours in air.

Chemical Stability of the G18/Crofer22 APU Interface

A recently developed SOFC interconnect alloy, Crofer22 APU [19], was investigated for its chemical compatibility with the sealing glass under different thermal histories. Figure 4 (a) shows the Crofer22 APU/G18/Crofer22 APU join after heat-treatment under the same thermal condition as for 446, i.e. in air at 850°C for one hour, followed by 750°C for 4 hours. The SEM images from the edge and interior areas on the cross-section of the joint are presented in Figure 4 (b) and (c), respectively. The areas marked as "a" in Figure 4 (b) and "b" in Figure 4 (c) are further enlarged in Figure 4 (d) and (e), respectively. It appeared that the microstructure of the Crofer22 APU/G18 interface was generally homogenous over the entire cross-section of the join. Even at the edge area, the sealing glass-ceramic was well-bonded to the stainless steel, and no gap and extensive interaction between the sealing glass and the steel occurred. In the interior, the sealing glass was adherent to the steel, and again no extensive interaction or porosity were observed along the Crofer22 APU/G18 interface. Thus, in comparison with 446, Crofer22 APU demonstrates an improved chemical compatibility with the sealing glass. This is due to the growth of a unique scale on Crofer22 APU during heat treatment, which is comprised of a $(Mn,Cr)_3O_4$ spinel top layer and chromia rich sub-layer [14,20]. The formation of a $(Mn,Cr)_3O_4$ spinel top layer mitigates the direct reaction of sealing glass with chromia and its vapor species.

To further examine the chemical stability between Crofer22 APU and G18, two Crofer22 APU coupons were joined with G18 (see Figure 5 (a)) in air at 850°C for one hour followed by heat treatment at 750°C for 24 hours (instead of the previous 4 hours). After the longer heat treatment, it was found that the glass-ceramic squeezed out during joining had turned light yellow. The sealing glass appeared to still be well bonded to the Crofer22 APU, as shown in Figure 5 (b) and Figure 5 (d) (which is an enlargement of the area marked as "a" in Figure 5(b)). Detailed analysis, however indicated that a detectable amount of barium chromate had formed both on the surface of the squeezed-out glass ceramic and along the interface at the edge area. Away from the edge area, porosity was evident in the interior of the joint, as shown in Figure 5 (c). The enlarged image in Figure 5 (e) further reveals the porosity and the microstructure of the Crofer22 APU/G18 interface. In a subsequent test, two Crofer22 APU coupons were joined using the sealant G18 at 850°C for one hour followed by heat treatment at 750°C for 168 hours (one week). Interfacial microstructures from the edge and interior areas of the join after this prolonged heat treatment are shown in Figure 6 (a), (b) and (c), respectively. Once again, the glass-ceramic squeezed out between the two Crofer22 APU coupons during joining turned yellow, indicating extensive formation of barium chromate. At the edge area, the sealing glass was separated from the ferritic stainless steel and barium chromate was also observable. In the interior, extensive porosity was found along the Crofer22 APU/G18 interface, though the sealing glass was still bonded to the stainless steel.

Thus, as evidenced by the prolonged test, Crofer22 APU, like the other ferritic stainless steels, reacts with the sealing glass to form barium chromate and interfacial porosity, although the kinetics of the interactions appear to be hindered due to the growth of a unique scale on that steel during heat treatment.

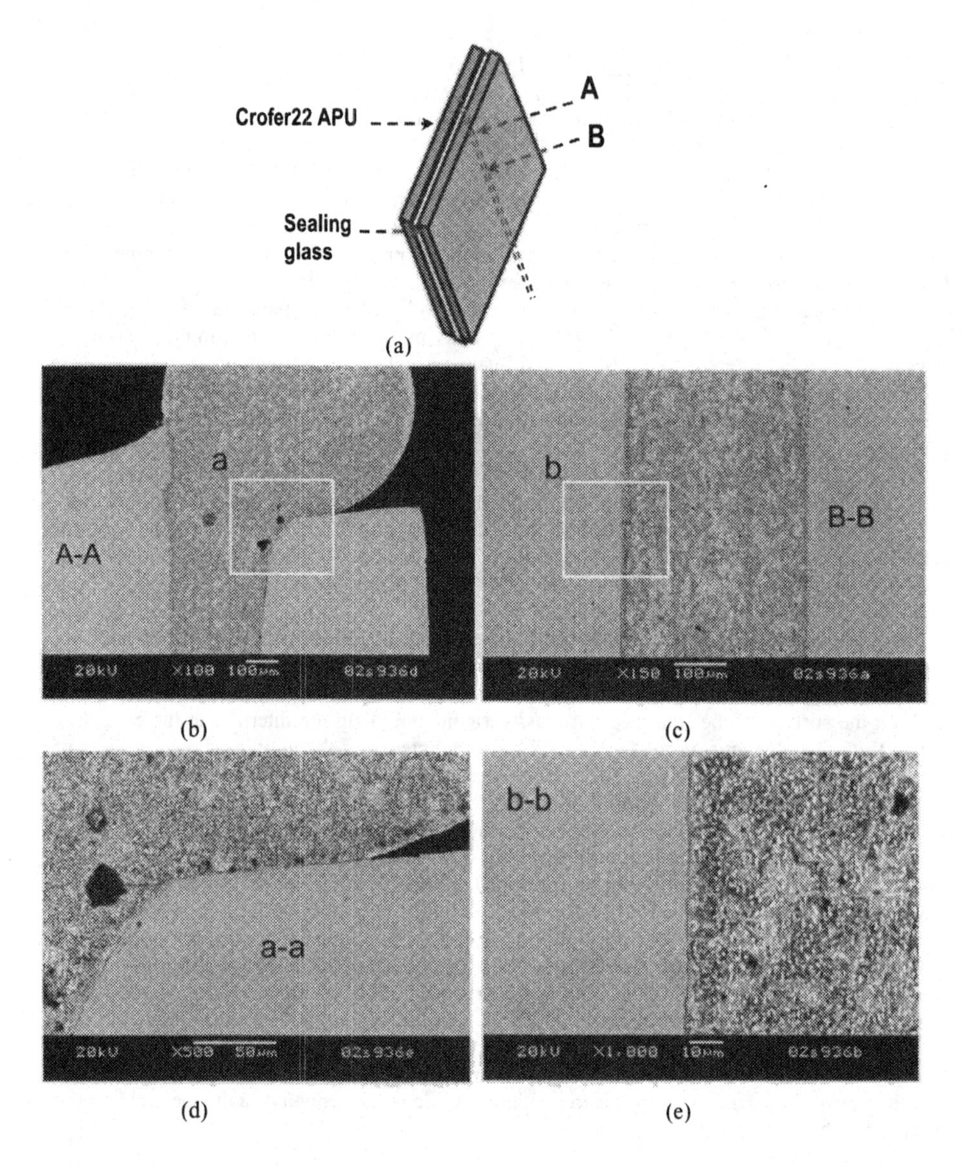

Figure 4. Interfacial reactions between G18 sealing glass and stainless steel Crofer22 APU: (a) a schematic of the joined "couple" (Crofer22 APU/G18/Crofer22 APU), and secondary electron SEM images of the interfacial cross-section at: (b) the edge area "A", (c) at the interior region B, (d) from the region marked as "a" in (b), and (e) from the region marked as "b" in (c). The Crofer22 APU coupons (12.7mm×12.7mm×1.0mm) were joined with G18 through heat treatment in air at 850°C for one hour, followed by 750°C for 4 hours.

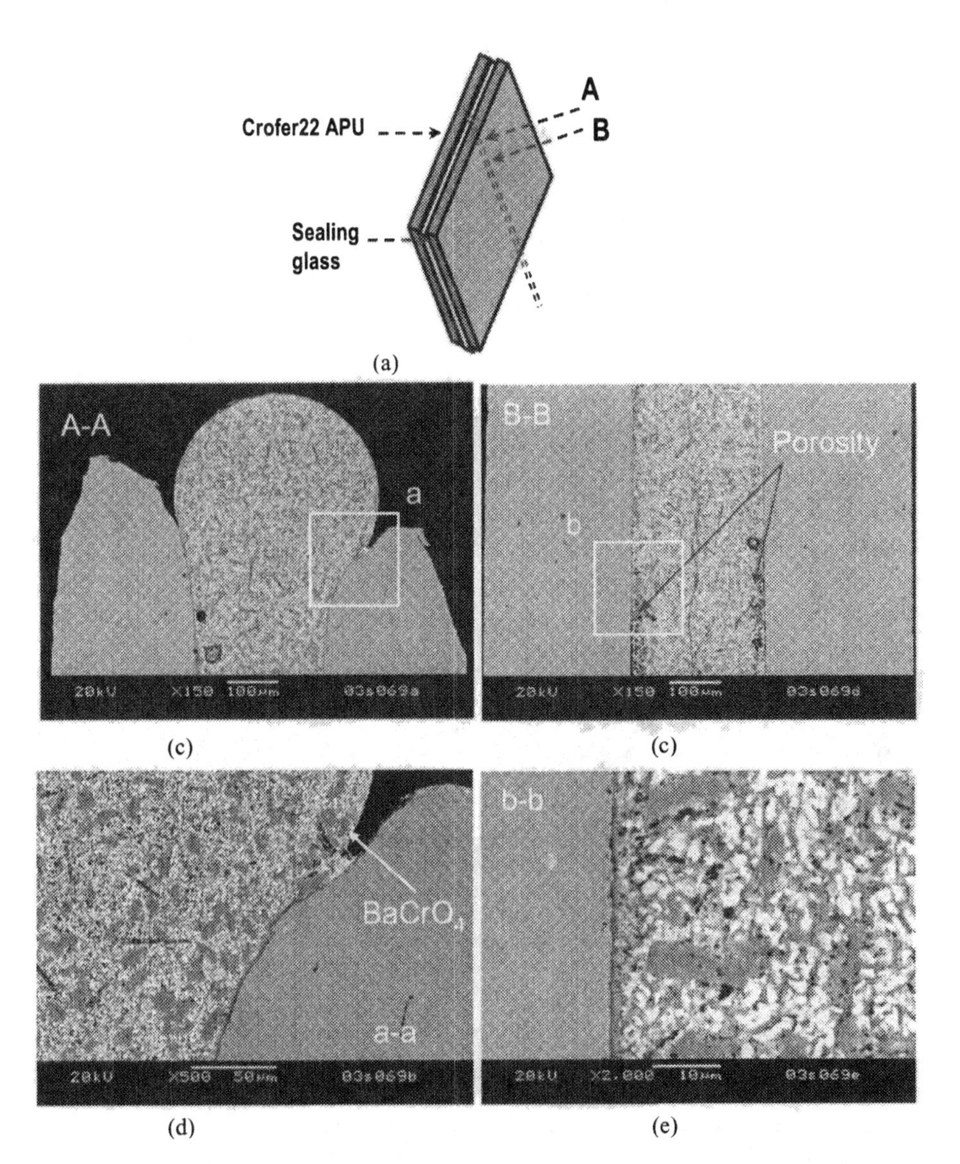

Figure 5. Interfacial reactions between G18 sealing glass and stainless steel Crofer22 APU: (a) a schematic of the joined "couple" (Crofer22 APU/G18/Crofer22 APU), and secondary electron SEM images of the interfacial cross-section at: (b) the edge area "A", (c) at the interior region B, (d) from the region marked as "a" in (b), and (e) from the region marked as "b" in (c). The Crofer22 APU coupons (12.7mm×12.7mm×1.0mm) were joined with G18 through heat treatment in air at 850°C for one hour, followed by 750°C for 24 hours.

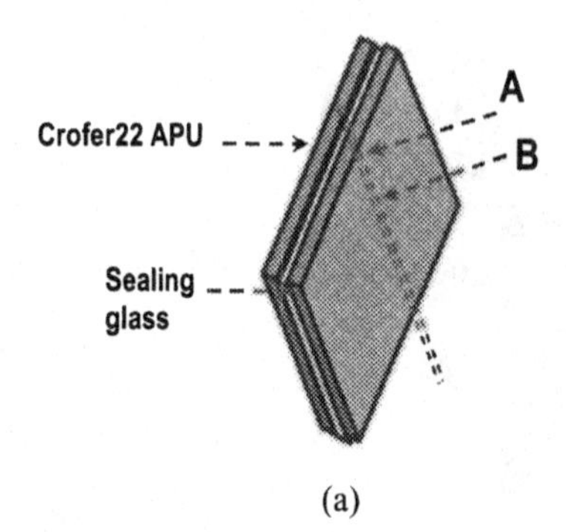

(a)

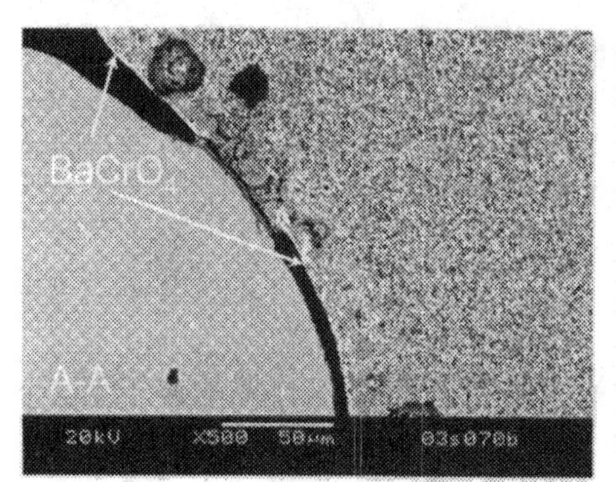

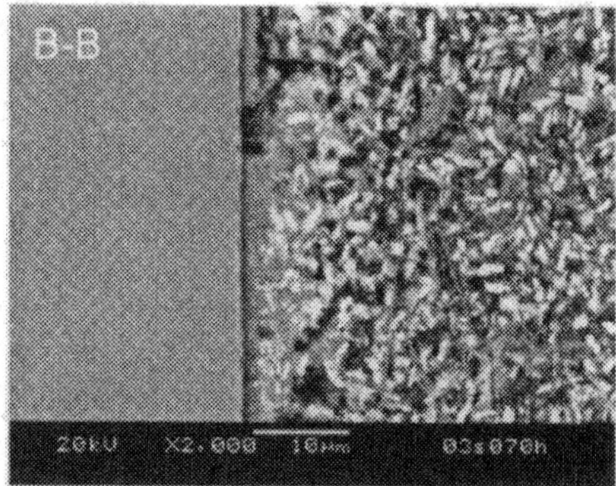

Figure 6. Interfacial reactions between G18 sealing glass and stainless steel Crofer22 APU: (a) a schematic of the joined "couple" (Crofer22 APU/G18/Crofer22 APU), and secondary electron SEM images of the interfacial cross-section at: (b) the edge area "A", and (c) at the interior region B. The Crofer22 APU coupons (12.7mm×12.7mm×1.0mm) were joined with G18 through heat treatment in air at 850°C for one hour, followed by 750°C for 168 hours (one week).

CONCLUSIONS

This work leads to the following conclusions:

1. The BCAS sealing glass-ceramic interacts with both the YSZ electrolyte and ferritic stainless steel interconnects. Overall, the YSZ electrolyte demonstrates good chemical compatibility and bonding to the glass-ceramic through limited reaction as well as good wetting that leads to the penetration of glass-ceramic into any available open porosity in the YSZ electrolyte. In contrast, the stainless steel interconnect reacts extensively with the sealing glass-ceramic resulting in an interface that is more prone to defects.
2. For traditional chromia forming stainless steels, the extent and nature of their interaction with the glass-ceramic depends on the exposure conditions and/or proximity of the

interface of sealing glass and ferritic stainless steel to the ambient air. At or near the edges, where oxygen from the air is accessible, the chromia scale grown on the steel and its vapor species react with BaO in the glass-ceramic, leading to the formation of $BaCrO_4$. In interior regions where the oxygen or air access is blocked, chromium or chromia dissolves into the BCAS sealing glass to form chromium rich solid solutions. The stainless steel also reacts with residual species in the sealing glass-ceramic to generate porosity in the glass-ceramic along the interface in the interior regions.

3. In comparison with traditional chromia forming stainless steels, the newly developed ferritic stainless steel Crofer22 APU exhibits improved chemical compatibility and bonding with the BCAS based-glass ceramics due to the growth of a unique scale on the alloy during high temperature exposures. Under prolonged heating, however, the alloy still visibly reacts with the sealing glass-ceramic, leading to the formation of barium chromate at the edge areas of the joins, solid solution phases, and porosity in the interior regions.

ACKNOWLEDGEMENTS

The authors would like to thank Nat Saenz, Shelly Carlson, and Jim Coleman for their assistance in metallographic and SEM sample preparation and analysis. The work summarized in this paper was funded as part of the Solid-State Energy Conversion Alliance (SECA) Core Technology Program by the U.S. Department of Energy's National Energy Technology Laboratory (NETL). PNNL is operated by Battelle Memorial Institute for the U.S. Department of Energy under Contract DE-AC06-76RLO 1830.

REFERENCES

1. B.C.H. Steele and A. Heinzel: "Materials for Fuel Cell Technologies," Nature, 2001, vol. 414, p. 345.
2. N.Q. Mihn: "Ceramic Fuel Cells," J. Am. Ceram. Soc., 1994, vol. 76, p. 563.
3. K. Huang, P.Y. Hou, J.B. Goodenough: "Characterization of iron-based alloy interconnects for reduced temperature solid oxide fuel cells," Solid State Ionics, 2000, vol. 129, p. 237.
4. W.J. Quadakkers, T. Malkow, J. Piron-Abellan, U. Flesch, V. Shemet, and L. Singheiser: in Proceedings of the 4th European Solid Oxide Fuel Cell Forum, A. McEvoy, Editor, 2000, vol.2, p. 827, the European SOFC Forum, Switzerland.
5. J. Piron-Abellan, V. Shemet, F. Tietz, L. Singheiser, and W.J. Quadakkers: in Proceedings of the 7th International Symposium on Solid Oxide Fuel Cells, H. Yokokawa and S.C. Singhal, Editors, 2001, vol. 2001-16, p. 811, The Electrochemical Proceedings Series, Pennington, NJ.
6. S.P.S. Badwal, R. Bolden, and K. Foger: in Proceedings of the 3rd European Solid Oxide Fuel Cell Forum, Ph. Stevens, Editor, 1998, vol. 1, p. 105, the European SOFC Forum, Switzerland.
7. T. Brylewski, M. Nanko, T. Maruyama, and K. Przybylski: "Application of Fe-16Cr Ferritic Alloy to Interconnector for A Solid Oxide Fuel Cell," Solid State Ionics, 2001, vol. 143, p. 131.
8. Z. Yang, K.S. Weil, D.M. Paxton, and J. W. Stevenson: "Selection and Evaluation of Heat Resistant Alloys for Solid Oxide Fuel Cell Interconnect Applications," J. Electrochem. Soc., 2003, vol. 150, pp. 1188-1201.

9. P. Kofstad: *Nonstoichometry, Diffusion and Electrical Conductivity in Binary Metal Oxides*, Wiley-Interscience, New York, 1972.
10. P. Kofstad and R. Bredesen: "High Temperature Corrosion in SOFC Environments," Solid State Ionics, 1992, vol. 52, p. 69.
11. K.D. Meinhardt, J.D. Vienna, T.R. Armstrong, and L.R. Peterson: "Glass-Ceramic Material and Method of Making," U.S. paten, No. 6,430,966, 2001.
12. Z. Yang, K.D. Meinhardt, and J.W. Stevenson: "Chemical Compatibility of Baruim-Calcium-Aluminosilicate-Based Sealing Glasses with Ferritic Stainless Steel Interconnects in SOFCs," J. Electrochem. Soc., 2003, vol. 150, pp. A1095-1101.
13. Z. Yang, J.W. Stevenson, and K.D. Meinhardt: "Chemical Interactions of Barium-Calcium-Aluminosilicate-Based Sealing Glasses with Oxidation Resistant Alloys," Solid State Ionics, 2003, vol. 160, pp. 213-225.
14. W.J. Quadadakkers, V. Shemet, and L. Lorenz: "Materials Used at High Temperatures for a Bipolar Plate of a Fuel Cell," U.S. patent, No. 2003059335, 2003.
15. Y. Matsuzaki and I. Yasuda: "Dependence of SOFC Cathode Degradation by Chromium-Containing Alloy on Compositions of Electrodes and Electrolytes," J. Electrochem. Soc., 2001, vol. 148, p. A126.
16. K. Hilpert, D. Das, M. Miller, D.H. Peck, and R. Weib: "Chromium Vapor Species over Solid Oxide Fuel Cell Interconnect Materials and Their Potential for Degradation Processes," J. Electrochem. Soc., 1996, vol. 143, p. 3642.
17. R. Weib, D. Peck, M. Miller, and K. Hillert: "Volatility of Chromium from Interconnect Material," in the Proceedings of the 17th Riso International Symposium on Materials: High Temperature Electrochemistry: Ceramics and Metals, F.W. Poulsen, N. Bonanos, S. Linderoth, M. Mogensen, and B. Zachau-Christianen, Editors, 1996, p.479, Denmark.
18. C.W.F.T. Pistorius and M.C. Pistorius: "Lattice Constants and Thermal-Expansion Properties of the Chromates and Selenates of Lead, Strontium and Barium," Z. Krist, 1962, vol. 117, p. 259.
19. W.J. Quadakkers, V. Shemet, and L. Singheiser, US Patent No. 2003059335.
20. Z. Yang, M.S. Walker, J. Hardy, G. Xia, and J.W. Stevenson: "Structure and Conductivity of Thermally Grown Scales on Ferritic Fe-Cr-Mn Steel for SOFC interconnect Applications," Journal of Electrochemical Society, 2004, in press.

Pd-MODIFIED REACTIVE AIR BRAZE FOR INCREASED MELTING TEMPERATURE

J.S. Hardy, K. S. Weil, and J. Y. Kim
Pacific Northwest National Laboratory
902 Battelle Blvd.
Richland, WA 99352

J. T. Darsell
Washington State University
PO Box 642920
Pullman, WA 99164-2920

ABSTRACT

Reactive air brazing (RAB) creates junctions between oxide surfaces that can withstand joining and operating temperatures up to 900°C in oxidizing environments. This makes RAB an attractive joining method for creating seals in solid oxide fuel cells (SOFCs) which operate at temperatures between 700 and 850°C and comprise cathode materials which are intolerant of reducing environments. Many planar SOFC designs require a two-step sealing process which in turn demands that the softening or melting point of the material used in the initial joining step be sufficiently higher than the joining temperature of the material used in the final step to ensure that the integrity of the initial seals is retained. The goal of this study is to investigate whether adding Pd to a Ag-Cu RAB might produce a braze composition that can be used in a two-step sealing process with an unmodified Ag-Cu RAB so that both sets of seals can be reactive air brazed. In order to model the initial sealing step in planar SOFCs which generally involves sealing a yttrium-stabilized zirconia (YSZ) electrolyte to a support frame, reactive air brazing of YSZ is the focus of this study. It was found that RAB compositions containing a 15 mol% Pd : 85 mol% Ag alloy with 8 mol% Cu added may provide the wettability, mechanical strength, and melting characteristics required for use in a two-step SOFC sealing approach.

INTRODUCTION

Reactive air brazing is a joining technique that is similar in function to active metal brazing. In both methods, a reactive component in the braze alloy modifies the faying surfaces of the parts to be joined in such a way that their interfacial energies with the remaining molten filler metal is decreased allowing for improved wetting. However, reactive air brazing differs from active metal brazing in that RAB can be carried out in air and produces a joint that is stable under oxidizing conditions at elevated temperatures making it amenable to joining oxide surfaces such as ceramics or high-temperature stainless steels that form stable oxide scales.

One alloy system that has been found to be well-suited for RAB is the Ag-Cu system in which the copper oxidizes in situ to form copper oxide. Meier et al.[1] reported that small additions of CuO greatly improve wetting of silver on alumina in inert atmospheres. Recognizing its potential for air-joining, Schüler et al.[2] demonstrated that 1 mol% CuO in Ag could indeed be used for joining alumina substrates in air. Reactive air brazing with Ag-Cu alloys containing various concentrations of Cu has since been studied on a number of oxide surfaces including alumina[3], yttrium-stabilized zirconia[4], lanthanum strontium cobalt ferrite[5], and pre-oxidized fecralloy[5].

Complex high temperature devices such as planar solid oxide fuel cell (pSOFC) stacks often require a two-step sealing process. For example, in pSOFC stacks, electroded YSZ electrolyte plates might be sealed into metallic support frames in one step. Then the frames supporting these fuel cell plates would be joined together in a separate sealing step to form the fuel cell stack. In this case, the initial seal should have a sufficiently high solidus temperature that it will not begin to remelt or lose integrity at the sealing temperature of the material used for the subsequent sealing step. Previous experience has indicated that Ag-CuO reactive air braze (RAB) sealing can be reliably performed at 970°C. Therefore, compositionally modifying the original Ag-CuO braze with Pd-additions such that the solidus temperature of the new braze is above 1025°C should provide two RAB compositions with a difference in melting points large enough to allow reactive air brazing of both sets of seals in the fuel cell stack. This study investigates the wettability of Pd-modified RAB compositions on 8% yttrium-doped zirconia (YSZ) and the mechanical strength of those brazes joining YSZ.

EXPERIMENTAL

The braze compositions investigated in this study were formulated on the basis of adding various concentrations of Cu to a 15 mol% Pd : 85 mol% Ag alloy, with the resulting brazes each having the general chemical formula, $(Ag_{0.85}Pd_{0.15})_{1-x}Cu_x$. Fine powders of silver (spherical, 0.5-1 μm average particle size, 99.9% pure by metals basis, Alpha Aesar, Ward Hill), copper (1-1.5 micron average particle size, 99% pure by metals basis, Alpha Aesar) and palladium (submicron, 99.9+% pure, Aldrich) were measured to the correct molar ratios and ball-milled in ethanol for 24 h. The resulting mixture was dried and sieved through a 200 mesh screen.

Differential scanning calorimetry (DSC) was performed on 2 mm dia. braze pellets pressed from approximately 5-10 mg of previously prepared powders using a Netzsch calorimeter (model STA 449C Jupiter) equipped with a high temperature furnace and a Type-S sample carrier. The experiments were performed in dry air flowing at a rate of 10ml/min with a heating rate of 10°C/min.

Sessile drop experiments were used to determine the contact angles of molten RAB pellets with varying concentrations of copper. The pellets had been uniaxially pressed from approximately 0.84 g of the braze powder mixtures in a 7 mm dia. die to form pellets approximately 7 mm in height. The pellets were placed upon the center of a 32 mm dia. plate of polycrystalline 8 mol% yttrium-doped zirconia (ZDY-8, CoorsTek) that had been polished to a <3 μm finish. The melting and wetting behavior of the braze pellets were recorded using a high speed video camera that was aligned to view the sample through a quartz window in a static air box furnace. Samples were heated to 900°C at 50°C/min at which point the heating rate was slowed to 10°C/min until melting was complete. At temperatures of 1150°C, 1200°C, and 1250°C, the temperature was held for 10 min to allow the molten braze to reach its equilibrium shape. Frames from the final seconds of each 10 min dwell were captured from the video camera and converted to computer images using VideoStudio6™ (Ulead Systems, Inc.) video editing software. These images were imported into Canvas™ (version 9.0.1, build 689; Deneba Systems, Inc.) graphics software which provided tools that facilitated measurement of the contact angles.

Four point bend tests measured the flexural strengths of brazed joints. For bend test specimens, rectangular YSZ bars were uniaxially pressed from 33 g of powder in a 32 mm by 64 mm die at 25 MPa and isostatically pressed at 135 MPa. The bars were then sintered at 1450°C for 1 hr to produce bars approximately 5.5 mm thick. The long, thin edges of the bars that were to be brazed together were polished to ensure a flat, smooth faying surface that is perpendicular to the length of the bend specimens. The polished edges of a pair of YSZ bars were brazed together at 1150°C for 30 min in air using a 70 wt% solids braze paste which was prepared by mixing a polymer binder (B75717, Ferro Corp.) with the appropriate amount of dry braze powder. Spring steel side clips and appropriately positioned refractory bricks were used to keep the bars properly aligned during firing. The joined bars were then ground to a 4 mm thickness and several 3 mm wide bend specimens were cut perpendicular to the brazed joint. Each individual bar was then chamfered before bend testing. The inner and outer spans used for four point bend testing were 20 and 40 mm respectively. A Bionix 400™ (MTS Systems Corp.) test stand applied the flexural load to the samples at a displacement rate of 0.5 mm/min and measured the load at failure. The failure stress was then calculated by TestWorks™ v3.10 (MTS Systems Corp.) materials testing software.

RESULTS AND DISCUSSION

DSC Measurements

Due to the increased materials cost associated with incorporating palladium into a braze alloy, initial DSC measurements were run to find a base alloy with a minimum Pd concentration in Ag that might elicit a sufficiently high melting point when Cu is added to it that subsequent brazing steps using unmodified Ag-Cu brazes might leave previously-fired Pd-Ag-Cu joints intact. As shown in table I, the liquidus and solidus temperatures obtained by DSC for Pd-Ag compositions agreed reasonably well with the phase diagram work by Karayaka and Thompson[6], coming within 10°C of the literature value in each case. Reactive air braze joints comprising Ag and Cu with no Pd can be reliably fired at 970°C. Therefore, it was decided that the experimentally measured solidus temperature of 1019°C for 10 mol% Pd in Ag may not provide a large enough buffer to ensure its integrity during a subsequent firing of a lower melting Ag-Cu RAB, especially taking into consideration the potential effects of Cu-additions on the melting characteristics of the Pd-Ag alloy. On the other hand, the 90°C buffer provided by adding 15 mol% Pd to Ag was deemed acceptable and this composition was selected as the base RAB alloy to which various amounts of Cu would be added for further investigations.

Table I. Experimental melting temperatures as measured by DSC are compared to literature values.

	DSC Melting Temperatures		Karayaka & Thompson	
Composition	Solidus (°C)	Liquidus (°C)	Solidus (°C)	Liquidus (°C)
10 mol% Pd in Ag	1019	1052	1030	1060
15 mol% Pd in Ag	1060	1103	1065	1100

The endothermic peaks measured by DSC that correspond to melting events in 15Pd-85Ag RAB alloys containing up to 8 mol% Cu are shown in fig. 1. A DSC curve for pure silver measured using the same heating rate and pellet configuration as for the RAB samples is also included for reference. During trial firings in a box furnace, samples with Cu-concentrations of 16 mol% or more reacted with the thin alumina crucibles used for DSC and were not measured in the calorimeter to avoid contamination of the instrument. Samples containing between 1 and 8 mol% Cu were each found to exhibit a small endotherm that is discernible at 983°C. This peak represents the solidus temperature for these compositions and corresponds to the onset of melting in a low melting minority phase. The relatively low intensity of the peak suggests that this phase must represent a very small fraction of the braze sample. The videotape of the sessile drop experiments observed this stage of melting as the formation of small, discrete beads of liquid on the surface of a continuous solid that made up the remainder of the pellet. Figure 2 comprises images captured from a video of a pellet heated through the various stages of melting detected by DSC. Because of the small volume fraction that appears to melt at 983°C and the persistence of a continuous solid phase, this stage of melting may not be detrimental to the integrity of a seal formed using these compositions. However, this must yet be confirmed experimentally. Once the minority phase has melted, the sample remains largely unchanged until the 1030-1040°C onset of a more intense melting event that overlaps with yet a third endothermic peak, indicating the melting of two remaining solid phases before the braze has reached its liquidus temperature of between 1074-1089°C. The specific temperatures at which selected melting events were detected for each composition by DSC are presented in table II.

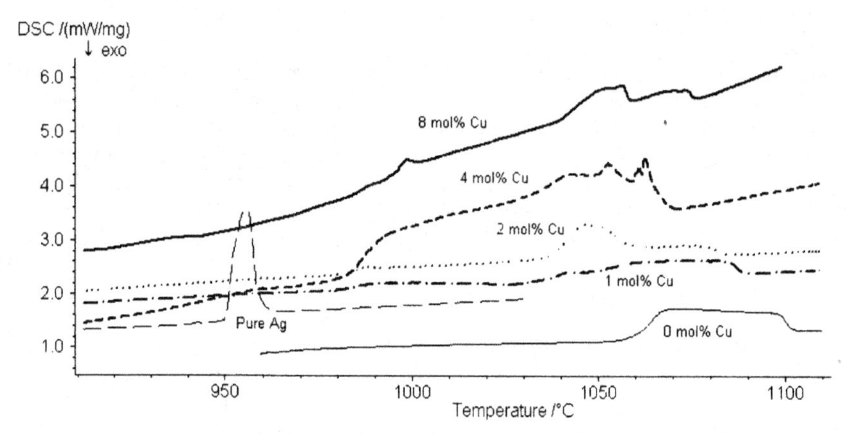

Figure 1 The endothermic peaks measured by DSC during heating at 10°C/min in air reveal the effects of Cu-content on the melting temperatures of 15Pd-85Ag reactive air brazes. The melting curve for pure Ag is included as a reference.

Table II. Effect of Cu-content on melting temperatures of 15Pd-85Ag alloys.

Cu-content (mol%)	Sweating Onset (°C)	Melting Onset (°C)	Melting Completion (°C)
0	n/a	1060	1103
1	983	1036	1089
2	983	1036	1084
4	983	1031	1074
8	983	1039	1076

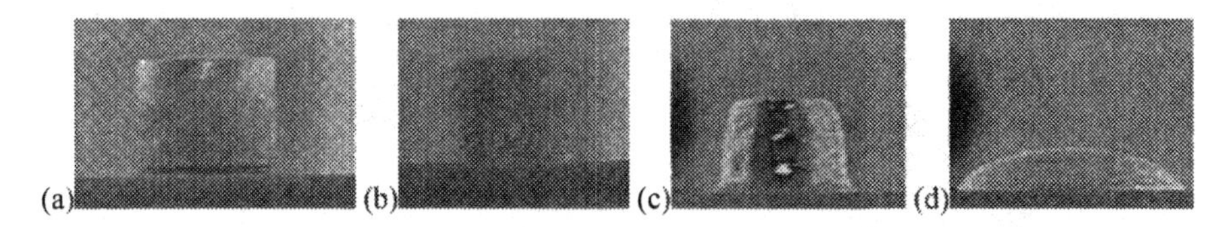

Figure 2 Captured video images showing the stages of melting: (a) solid pellet, (b) minority phase melts forming discrete beads of liquid, (c) onset of second melting stage, and (d) completion of melting.

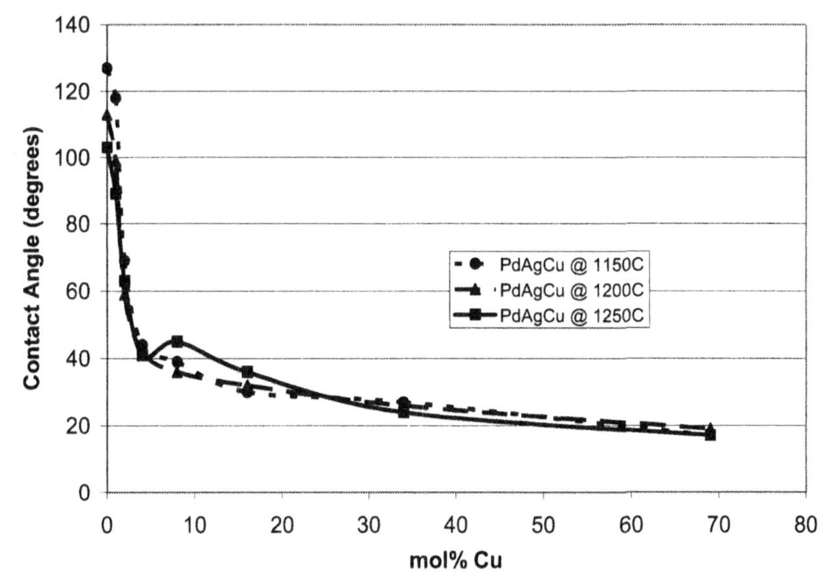

Figure 3 Contact angle is plotted as a function of copper additions in 15Pd-85Ag alloy at three different temperatures.

Sessile Drop Experiments

The contact angles of 15Pd-85Ag alloys shown in fig. 3 decrease with increasing Cu-additions indicating improved wettability on YSZ. Notwithstanding the dramatic decrease in contact angle observed when 1 mol% Cu is added to the 15Pd-85Ag alloy, non-wetting, which is generally defined as an obtuse contact angle, was still observed in samples with 1 mol% Cu added. Wetting was first observed in samples with 2 mol% Cu and continued to improve rapidly up to 4 mol% Cu. Further increases in Cu-content only resulted in modest decreases in the contact angle. During the tests, all compositions reached their equilibrium shapes almost immediately upon reaching the measurement temperatures, never taking more than 60 seconds to equilibrate suggesting that the interfacial reactions probably occur soon after melting allows the copper oxide to migrate to the YSZ surface where it reacts to produce an interface that is more energetically favorable for wetting by the molten filler metal. No consistent relationship was observed between contact angle and temperature.

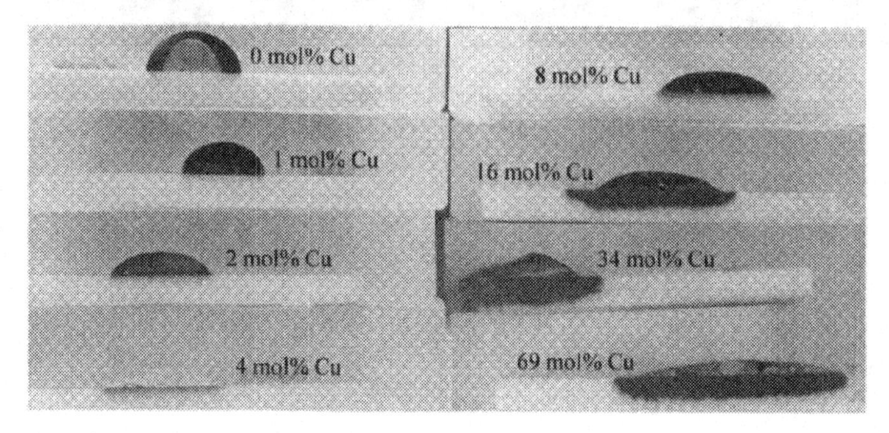

Figure 4 After sessile drop testing, samples were sectioned to reveal diffusion of the copper oxide into the YSZ substrate creating a reaction band in some samples.

After sessile drop tests were completed, a distinct black ring was observed surrounding the braze beads containing at least 16 mol% Cu. The photograph of sectioned pellets in fig. 4 shows that the interfacial reaction with copper oxide extended deep within the YSZ substrate in samples with more than 8 mol% Cu. Although much less obvious than that seen in samples with higher Cu-concentrations, close inspection reveals a thin, faint region of discoloration that is also present in the YSZ directly beneath the 2, 4, and 8 mol% Cu braze beads. The discoloration becomes increasingly prominent with increasing Cu-content.

Four Point Bend Tests

The chart in fig. 5 shows the effect of Cu-content on the average flexural strength of RAB joints between YSZ bars as measured using the four point bend method. Although Pd-containing reactive air braze joints were found to be weaker than joints formed using Ag-Cu brazes containing no Pd at the four Cu-contents for which comparisons were available, the trend of average flexural strength was strikingly similar. At three of the four Cu-contents compared, the Pd braze joints showed average flexural strengths that were between 11 and 12 MPa weaker than the analogous joints containing no Pd. In both braze families, compositions containing between 4 and 8 mol% Cu provided bonds with YSZ surfaces that were stronger than other copper contents. This is believed to be due to a balance that exists in this concentration range between wettability which improves with increasing Cu-content and contributes to a stronger bond, and an increasing amount of Cu-oxide in the fired braze which is brittle and detracts from the strength of the joint.

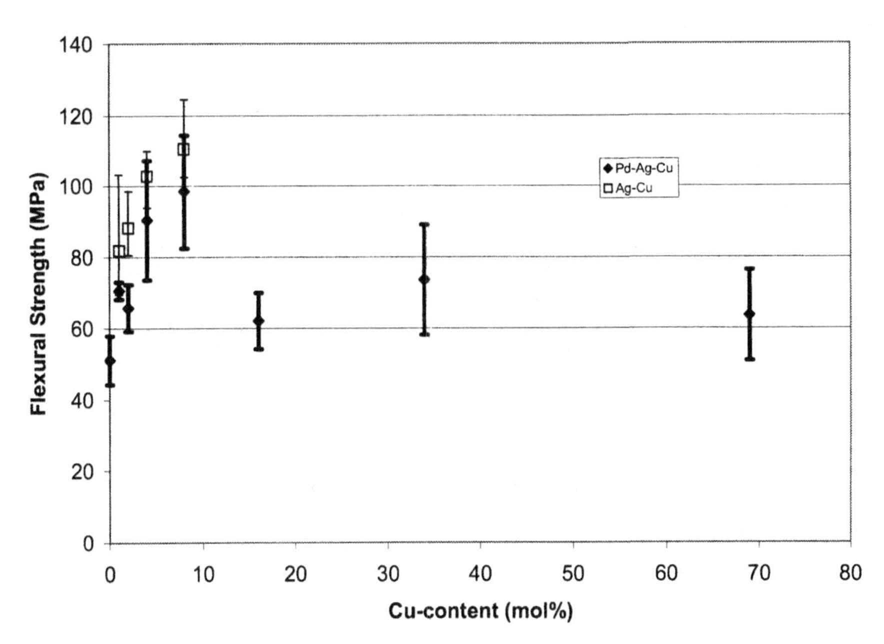

Figure 5 Average four point bend strengths of YSZ bars joined using 15Pd-85Ag alloys with varying amounts of Cu added are compared to values measured for brazes containing no Pd.

CONCLUSIONS

Based solely on melting characteristics, a 15 mol% Pd : 85 mol% Ag alloy which melts at 1060°C would be amenable to forming the initial junction in a two-step RAB joining process with Ag-Cu braze used for the final joining step. However, Cu must be added in order to improve wettability on YSZ and thereby increase the probability of forming strong, hermetic seals. It was found that samples containing between 1 and 8 mol% Cu additions give rise to a low melting phase that is apparently present in small concentrations and begins to melt at 983°C to form discrete beads of liquid that begin to appear on the surface of the braze. However, full-scale melting of the remainder of the Pd-modified RAB does not begin until 1030 to 1040°C. Therefore, depending on the effect the presence of localized regions of low melting phase has on the integrity of the Pd-braze seal during a subsequent Ag-Cu braze firing at 970°C, these compositions might be compatible with Ag-Cu RAB. The mechanical strength of Pd-containing braze joints to YSZ were weaker than analogous RAB compositions containing no Pd by 11 to 12 MPa at three of the four Cu-concentrations compared. However, samples containing 4 and 8 mol% Cu nonetheless exhibited average flexural strengths of greater than 90 MPa. In view of the melting characteristics and flexural strengths results, $(Ag_{0.85}Pd_{0.15})_{0.92}Cu_{0.08}$ appears to be the most promising Pd-RAB examined in this study.

ACKNOWLEDGMENTS

The authors would like to thank Nat Saenz, Shelly Carlson and Jim Coleman for their assistance in preparation of metallographic samples and SEM analysis work. This work was supported by the U. S. Department of Energy, Office of Fossil Energy, Advanced Research and Technology Development Program.

REFERENCES

[1]A.M. Meier, P.R. Chidambaram, and G.R. Edwards, "A Comparison of the Wettability of Copper-Copper Oxide and Silver-Copper Oxide on Polycrystalline Alumina," *Journal of Materials Science*, **30** [19] 4781-6 (1995).

[2]C.C. Shüler, A. Stuck, N. Beck, H. Keser, and U. Täck, "Direct Silver Bonding - An Alternative for Substrates in Power Semiconductor Packaging," *Journal of Materials Science: Materials in Electronics*, **11** [3] 389-96 (2000).

[3]J.Y. Kim, J.S. Hardy, and K.S. Weil, "Effects of CuO content on wetting behavior and mechanical properties of Ag-CuO braze for Ceramic Joining", *J. Am. Ceram. Soc.*, in review.

[4]J. Y. Kim, K. S. Weil, and J.S. Hardy, "Wetting and Mechanical Characteristics of the Reactive Air Braze for Yttria-Stabilized Zirconia (YSZ) Joining," this publication.

[5]K.S. Weil and J.S. Hardy , "Brazing a Mixed Ionic/Electronic Conductor to an Oxidation Resistant Metal," *Advances in Joining of Ceramics, Ceramic Transactions*, Volume 138

[6]I. Karayaka and W.T. Thompson, *Bull. Alloy Phase Diagrams*, **9** [3] 237-243 (1988).

EVALUATION OF GOLD ABA BRAZE FOR JOINING HIGH TEMPERATURE ELECTROCHEMICAL DEVICE COMPONENTS

K. Scott Weil
Pacific Northwest National Laboratory
P.O. Box 999
Richland, WA 99352

Joseph P. Rice
School of Mechanical and Materials Engineering
Washington State University
Pullman, WA 99164

ABSTRACT

The oxidation behavior of a commercially available ceramic-to-metal braze alloy, Gold ABA, was investigated to evaluate its potential use in joining solid-state electrochemical device components. High temperature air exposure studies were performed on as-received braze alloy foils, on braze couples prepared with yttria-stabilized zirconia (YSZ) and 430 stainless steel substrates, and on brazed YSZ/430 joints. The results of our investigations demonstrate that the substrates can play an important role in determining the overall oxidation resistance of the brazed joint.

INTRODUCTION

The development of high-temperature electrochemical devices such as oxygen and hydrogen separators, fuel gas reformers, solid oxide fuel cells, and chemical sensors is part of a rapidly expanding segment of the solid state technology market [1]. These devices employ an ionic conducting ceramic membrane that establishes an electrochemical potential either under a voltage gradient (which can be used to carry out elemental separations, such as oxygen from air) or under a chemical gradient (in order to develop an electrical potential and thereby generate electrical power, as occurs in solid oxide fuel cells). Because the device operates under an ionic gradient that develops across the electrolyte, hermiticity across this layer is paramount. Not only must this thin ceramic membrane be dense with no interconnected porosity, but it must be connected to the rest of the device, typically constructed from a heat resistant alloy, with a high-temperature, gas-tight seal. A significant engineering challenge in fabricating these devices is how to effectively join the thin electrochemically active membrane to the metallic body of the device such that the resulting seal is hermetic, rugged, and stable during continuous high temperature use over its operational lifetime.

The leading ceramic-metal materials set for these applications is the solid-state electrolyte, YSZ, and ferritic stainless steel. Historically high temperature glasses, such as those from the borate- or phosphate-doped aluminosilicate families, have been used to join these materials in commercial electrochemical devices [2]. However, metal alloy brazes promise better performance with respect to thermal shock, mechanical shock, and thermal cycling, key properties for the use of these devices in military and commercial transportation applications. An example of where ceramic-metal joining is required in an electrochemical device is illustrated in Figure 1, which displays a schematic drawing of a planar oxygen generator. Shown in the inset is a depiction of how oxygen transport takes place in the device. Also noted are the conditions under which the O_2 generator, and therefore the YSZ/metal joint, is expected to operate. The device consists of a stack of individually sealed, serially connected YSZ membranes, which under an imposed voltage catalytically ionize oxygen from an incoming air stream on one side and transport the resulting ions up a chemcial gradient to the other. To generate a sufficient rate of ionic transport within the stack, it must be operated at high temperature, normally on the order of 500 – 1000°C. Unfortunately, this temperature is also high enough for thermally activated processes to take place which may be deleterious to device performance, such as interdiffusion, oxidation, and interfacial solid state reaction - particularly over the expected lifetimes of these devices, estimated to be 3000 – 30,000+ hours depending on the specific application. Thus, the joint must be functionally stable and hermetic over long periods of time at operating temperature and be capable of surviving numerous thermal cycles from operating temperature down to room temperature and back up.

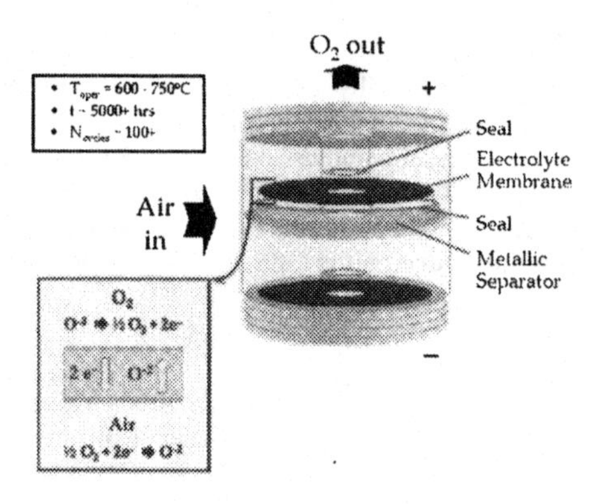

Figure 1 A schematic drawing of a planar oxygen generator.

Before a given braze alloy can be considered for high temperature application in air, its oxidation behavior and that of the as-brazed joint must be thoroughly characterized. Previous oxidation studies on active metal brazes have focused primarily on copper-silver based alloys, typically those bearing a small amount of titanium as the active metal species. In general it was found that these particular alloys are not capable of surviving sustained exposure to air at temperatures above 500 - 600°C [3-5]. Little work has been conducted on high temperature brazes or on the effects of the adjoining metal and/or ceramic substrates. As part of a broader investigation into the suitability of commercial active metal brazes for electrochemical device application, we have investigated the oxidation properties of the gold-based active metal braze, Gold ABA. This paper presents recent findings from our studies on both the high temperature oxidation behavior of Gold ABA and how the oxidation properties change due to the adjoining YSZ and 430 stainless steel substrates.

EXPERIMENTAL PROCEDURE

Materials and Sample Preparation

The braze alloy employed in these experiments was Gold ABA, which has a composition of Au-3Ni-0.6Ti. The braze was supplied as 0.150 mm thick foil from Wesgo Metals Inc. Samples measuring ½cm square were used for brazing. 5% yttria stabilized zirconia (5YSZ) and thin gauge 430 stainless steel (430 SS - Fe, 18% Cr, >1% Si) were selected as the model ceramic electrolyte membrane/structural metal system for this study. 5YSZ electrolyte substrates were fabricated by using tape casting and sintering procedure developed at Pacific Northwest National Laboratory. 5YSZ powder (Zirconia Sales, Inc.) was ball milled with a proprietary binder and dispersant system in 2-butanone/ethyl alcohol for 2 days to form a slurry. The slurry was cast onto silicone-coated mylar by the doctor blade technique to form green tapes with an as-dry thickness of approximately 200μm. Thicker tapes were prepared by laminating two of the green tapes at 175°C and 500psi for 2 minutes. Discs measuring 25mm in diameter were cut from this thick tape using a circular hot knife. The green discs were then sintered at 1400°C for 1 hour in air and creep flattened between two ground flat alumina plates at a temperature of 1450°C for two hours. The final diameter and thickness of the YSZ substrate discs were nominally 18mm and 200μm, respectively.

430 stainless steel (430 SS) was chosen as the model representative for the device housing material because: (1) its coefficient of thermal expansion (CTE) most closely matches that of the YSZ, (2) it displays good oxidation properties at high temperature, and (3) it is the most widely produced ferritic stainless product, making it a strong low-cost candidate for commercial high temperature electrochemical devices. The as-received ~½ mm (20 mil) sheet of 430 was sheared into 2 cm x 2 cm square samples and polished on both sides with 1200 grit SiC paper. The samples were flushed with de-ionized water to remove the grit and ultrasonically cleaned in acetone for 10 minutes.

Testing and Characterization

Our objectives in this study were to understand the mechanisms by which oxidation occurs in Gold ABA and what role the substrates might have in modifying this oxidation behavior. As shown in the flow chart in Figure 2, the braze alloy was exposure tested in four different configurations: (1) the braze foil alone, (2) a braze/5YSZ couple, (3) a braze/430 SS couple, and (4) an 5YSZ/braze/430 SS joint. For comparison, exposure testing was conducted in both high vacuum ($<10^{-6}$ torr) and in flowing air at 800°C for 50, 100, and 200 hours. Heating and cooling rates of 5°C/min were employed during testing. Changes in the composition and microstructure of these samples were compared against the baseline microstructures of the alloy in the unoxidized state and of the brazed couples and joints in the as-joined condition. Prior to joining and testing, all materials were thoroughly cleaned and degreased in acetone and ethanol. The

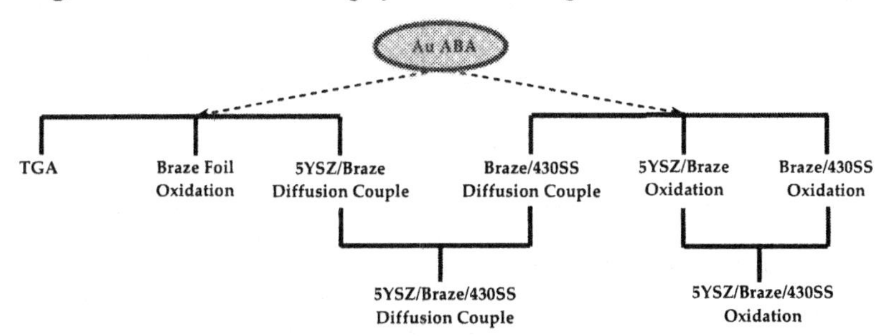

Figure 2 Flow chart of the oxidation experiments.

wetting and joining samples were prepared by heat treating under 10^{-6} torr vacuum at 3°C/min to 1080°C, holding at 1080°C for ½ hr, and cooling at 3°C/min. Microstructural analyses of the various vacuum treated and oxidized specimens were conducted on as-polished cross-sectioned samples using a JEOL JSM-5900LV scanning electron microscope (SEM) equipped with an Oxford energy dispersive X-ray analysis (EDX) system.

RESULTS AND DISCUSSION

Shown in Figures 3(a) and (b) are micrographs of the braze microstructure in the as-received condition. The cross-sectional microstructure of the as-received Au ABA foil consists of fine, micron-sized equiaxed Ni/Ti-rich precipitates randomly dispersed in a Au-rich matrix. The precipitates are composed of ~60 a/o gold and 20 a/o each of nickel and titanium, whereas the matrix contains ~90 a/o gold and 8 and >0.5a/o nickel and titanium, respectively. The microstructure near the surface of the as-received Au ABA sample was found to be approximately the same as that in the center.

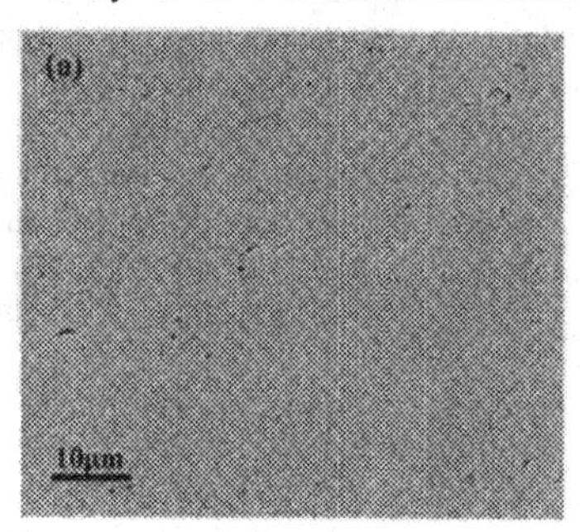

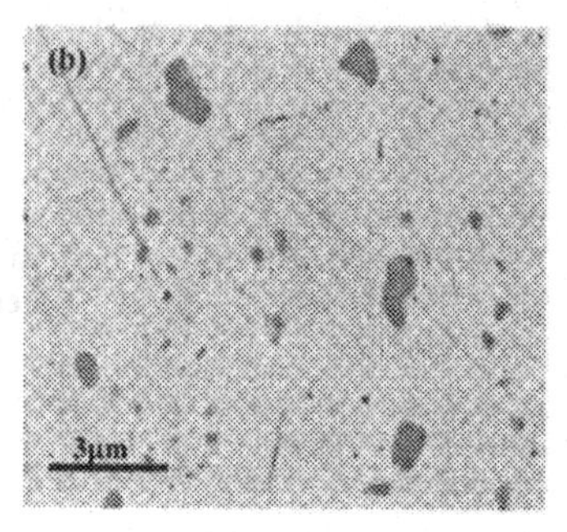

Figure 3 Cross-sectional SEM micrographs of Au ABA braze foils in the as-received condition: (a) 700x and (b) 5000x maginification.

The microstructures of the Au ABA/5YSZ, Au ABA/430 SS, and 5YSZ/Au ABA/430 SS specimens are shown in Figures 4(a) – (c). As seen in Figure 4(a), Au ABA exhibits good wetting and adherence with 5YSZ. In comparison with the as-received foil, the microstructure of this specimen contains virtually no precipitates, appearing as a homogeneous alloy up to the interface with the ceramic substrate. EDX measurements taken at several regions within the bulk of the braze and near the braze/5YSZ interface indicate that the alloy is composed, on average, of ~97 a/o Au, 2.5 a/o Ni, and >0.5 a/o Ti. Along the braze/5YSZ interface is a dense, ½ - 1μm thick reaction zone containing essentially a solid solution of titanium oxide and zirconium oxide. Local chemical analysis suggests that the metal species in this region are fully oxidized.

The microstructure of the braze in the Au ABA/430 SS couple, Figure 4(b), is quite different from that of the foil and the braze in the Au ABA/5YSZ couple. The primary reason for this is the partial dissolution of the underlying stainless steel substrate into the molten braze during joining, which subsequently establishes not only a new microstructure and composition in the bulk braze filler region [on the left side of Figure 4(b)], but also a mixing zone between the braze and the substrate. In the joining specimen in Figure 4(c), it is this diffusional mixing zone that is responsible for the observed differences in the braze/5YSZ interface relative to the braze-ceramic couple in Figure 4(b). Note that in the 5YSZ/Au ABA/430 SS joining sample, no continuous mixed oxide reaction zone is found adjacent to the 5YSZ faying surface. Instead much of the titanium appers to be tied up as fine scale, equiaxed TiO_2 particles in the braze matrix, seen as small black regions in the braze microstructure. The braze side of the 5YSZ/Au ABA interface consists primarily of two phases, iron-nickel rich precipitates in a gold-based matrix, with both phases containing measurable amounts of chromium. Also apparent in the microstructure of the braze filler are needle-shaped Widmanstätten type iron-titanium precipitates.

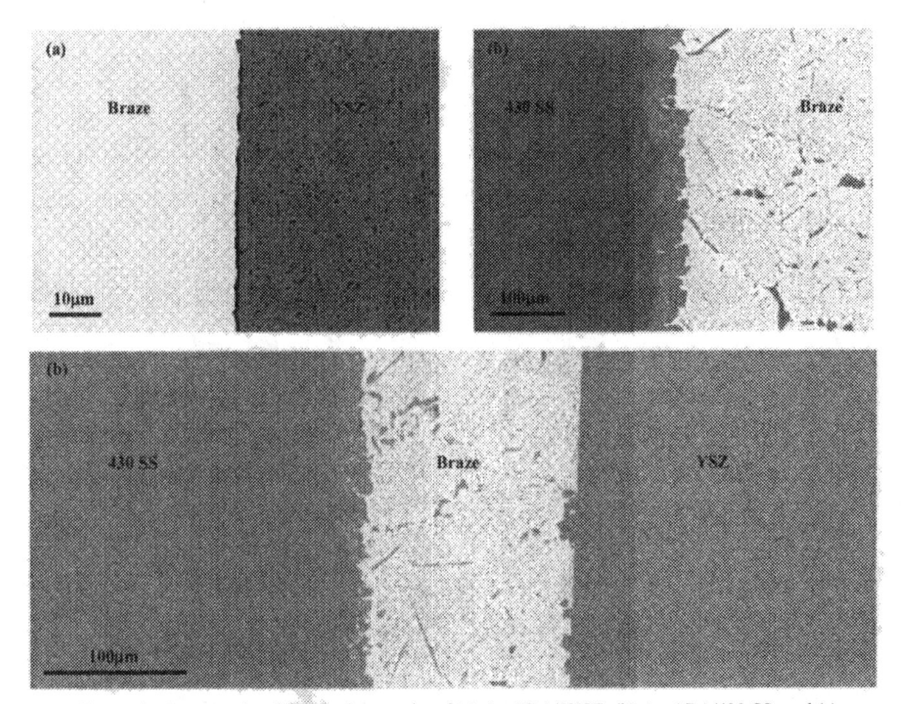

Figure 4 Cross-sectional SEM micrographs of (a) Au ABA/5YSZ, (b) Au ABA/430 SS, and (c) 5YSZ/Au ABA/430 SS specimens in the as-joined condition.

The corresponding microstructures for the full array of specimens aged under vacuum at 800°C are shown in Figures 5(a) – (l). In the braze foil specimens, we observed that precipitate coarsening occurs as a function of time at temperature. The original Ni/Ti-rich precipitates grow from ~1μm on average in the as-received foil to ~2μm in the sample annealed for 100hrs. Conversely, the microstructures of the Au ABA/YSZ and the Au ABA/430 SS couples displays little change as a function of annealing time. The reaction zones of the ceramic couple remains unchanged in thickness or composition, while the diffusional mixing zone of the brazed 430 coupons show a small amount of precipitate coarsening. Similarly, the microstructures of the joining specimens evolve only slightly during annealing. The Widmanstätten precipitates thicken to a small degree, their morphology becoming more blocky in appearance at longer annealing times. The composition however remains essentially unchanged both within the precipitates as well as in the surrounding matrix. Also note that annealing appears to have little effect on the appearance of the Au ABA/5YSZ interface.

Figures 6(a) – (l) show the effect of exposure under oxidizing conditions in each of the different specimen types. In all three braze foil specimens, Figures 6(a) – (c), oxidation is aggressive with a thick NiO scale layer (dark gray phase) forming at the exposed surfaces, as well as penetrating along grain boundaries well into the alloy. Scale penetration becomes more extensive the longer the foil is exposed to the high-temperature oxidizing condition. Adjacent to a majority of the scale in each specimen are regions within the braze containing small voids. The

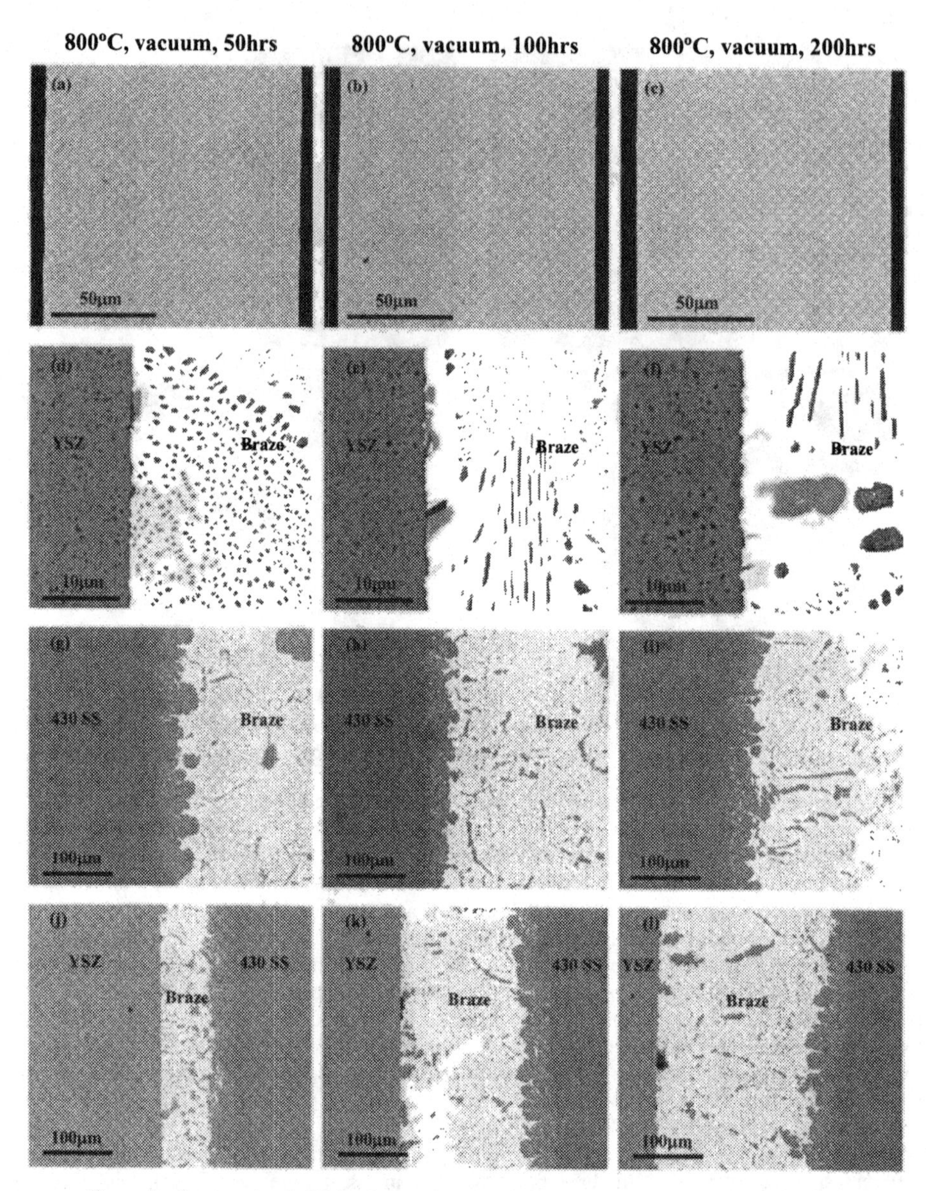

Figure 5 Cross-sectional SEM micrographs of the following specimen configurations heat treated under vacuum at 800°C for the time periods specified in the figure: (a) – (c) Au ABA foil, (d) – (f) Au ABA/5YSZ, (g) – (i) Au ABA/430 SS, and (j) – (l) 5YSZ/Au ABA/430 SS.

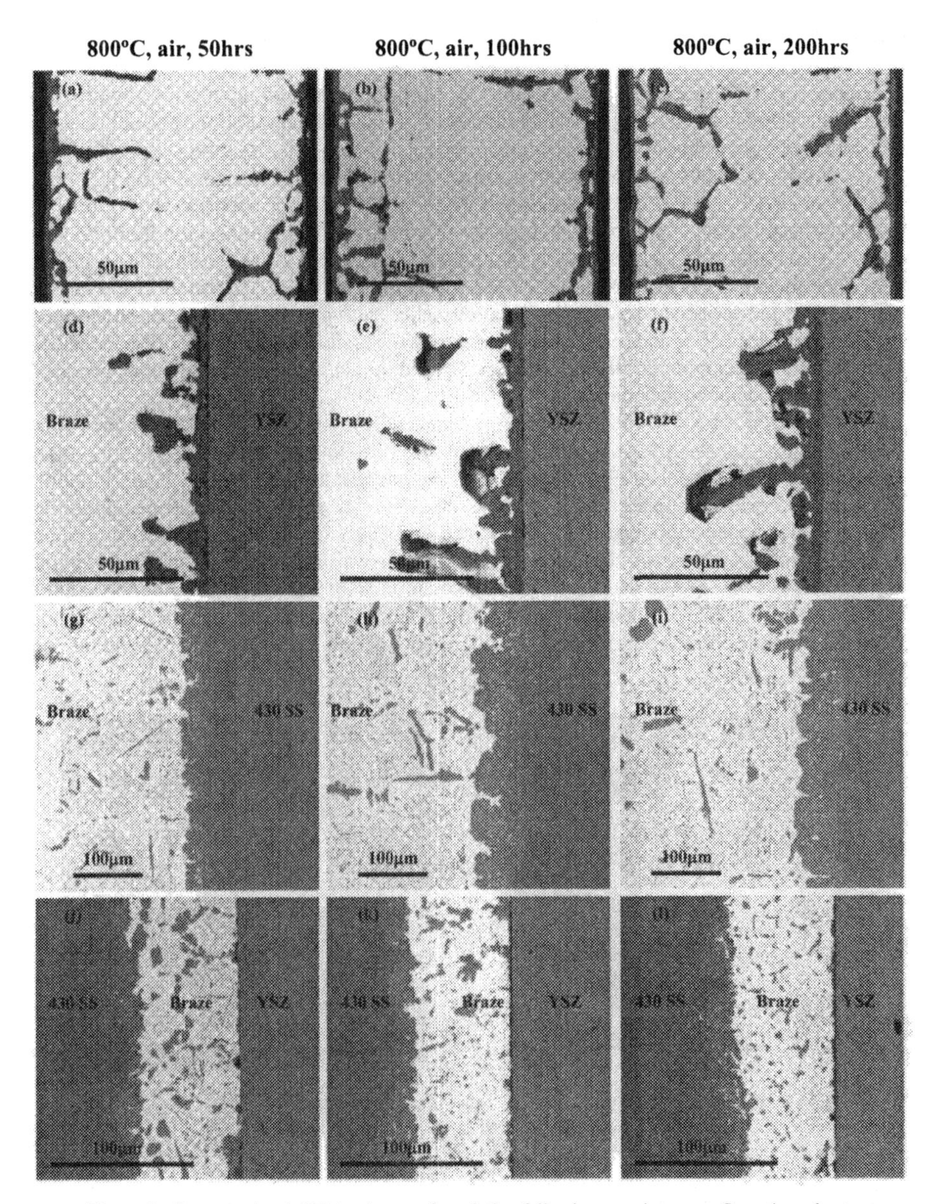

Figure 6 Cross-sectional SEM micrographs of the following specimen configurations heat treated in air at 800°C for the time periods specified in the figure: (a) – (c) Au ABA foil, (d) – (f) Au ABA/5YSZ, (g) – (i) Au ABA/430 SS, and (j) – (l) 5YSZ/Au ABA/430 SS.

voids are likely generated as the result of mass transport of nickel at the braze/scale interface through the oxide layer at a rate which is greater than the inward movement of oxygen; a typical mechanism for NiO scale formation in non-Cr, Si, and Al containing nickel alloys [6]. In addition, the samples are generally devoid of the Au-Ni-Ti precipitates originally found in the as-received braze, indicating that the alloying elements completely oxidized during exposure testing, confirming results from previously reported thermogravimmetric analysis data [7].

In the oxidized Au ABA/5YSZ specimens, Figures 6(d) – (f), a NiO sub-layer forms just beneath the reaction zone and extends along grain boundaries, penetrating into the braze filler material. Again a layer of fine porosity is observed adjacent to the NiO, presumably originating in the same manner as that found in braze foil oxidation, i.e. due a differential in the rates of mass transport between the two reacting species, nickel and oxygen. The thickness of the NiO scale and the extent of oxidation damage increases with prolonged exposure. After remaining 200hrs in high-temperature air, the internal porosity within the filler is substantial and the couple exhibits signs of incipient delamination.

In the oxidized Au ABA/430 SS specimens, Figures 6(g) – (i), the general appearance of the interfacial region, i.e. the diffusional zone, is nearly identical to those found in the corresponding vacuum annealed specimens in Figures 5(g) – (i). The most significant change occurs at the free surface on the braze-side of the couple, Figures 7(a) and (b), which can be compared with the oxidized foil specimens in Figures 6(a) – (c). Like the foil specimens, an oxide scale forms along the exposed surface. However, the scale is much thinner and more adherent to the underlying alloy and consists of Cr_2O_3 doped with a small amount of TiO_2. The scale grows as a function of exposure time, from 2.2μm in the 50hr specimen to 5.5μm in the 200hr specimen. The source of the chromium appears to be the matrix, which exhibits a 1μm thick chromium depletion zone adjacent to the scale.

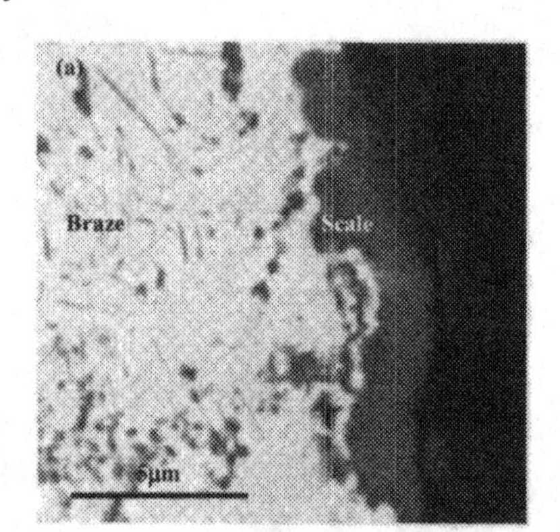

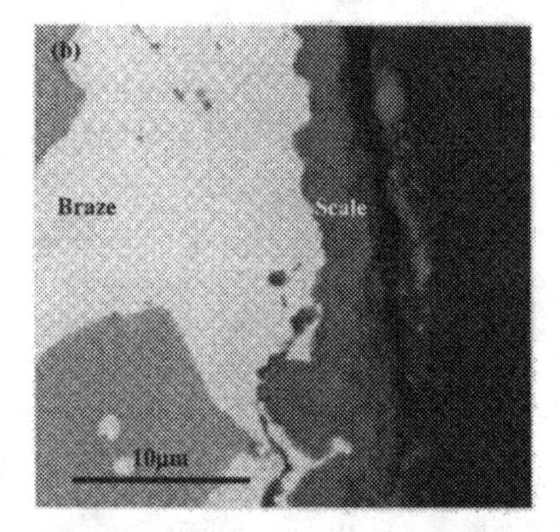

Figure 7 Cross-sectional SEM micrographs of the free surfaces on the braze side of two of the Au ABA/430 SS specimens oxidized at 800°C for: (a) 50hrs and (c) 200hrs.

The combined effects of the 5YSZ and the 430 SS substrates on Au ABA braze joint oxidation can be seen by comparing the micrographs of the oxidized specimens, Figures 6(j) – (l), with those of the previous samples. For example, in comparison with the corresponding vacuum annealed specimens, an oxidation zone adjacent to the YSZ faying surface is apparent in the air exposed joints. Composed of chromia, this layer is fully continuous and measures ~ 1μm thick after 50hrs of oxidation and grows to ~3μm after 200hrs. As was found in the oxidized Au ABA/430 SS specimens, the source of the chromium appears to be the braze filler matrix as a Cr-depletion zone arises adjacent to the oxide layer. Note also in Figures 6(j) – (l) that the braze microstructure also evolves. The needle-shaped iron-titanium precipitates in the Cr-depletion zone have shrunken in size compared to the vacuum annealed specimens. In addition a new precipitate has appeared, also with an elongated morphology but a lower c/a ratio, which forms in random directions as opposed to perpendicular crystallographic directions of the Widmanstätten precipitates. EDX analysis of the precipitates suggests that they are composed of iron, nickel, and oxygen although a quantitative ratio of these elements could not be obtained because of their size. The formation of the Cr_2O_3 is particularly striking when compared to the oxidized Au ABA/5YSZ couples. The chromia appears to protect the joint from the aggressive oxidation, particularly intergranular oxidation, observed in Figures 6(d) – (f).

CONCLUSIONS

The effect of substrate composition on the oxidation behavior of Au ABA at 800°C was examined by comparing samples annealed/oxidized for 50 – 200hrs. Our observations from these tests can be summarized as follows:

(1) When exposed to 800°C air, the braze foil displays aggressive oxidation, with nickel and titanium oxide forming along the grain boundaries well into the alloy. Scale formation was found to be accommodated by internal porosity and it was observed that the nickel and titanium alloying agents were completely oxidized.

(2) In the Au ABA/5YSZ couples, the braze forms a TiO_2/ZrO_2 reaction zone along the interface with the electrolyte. Subsequent long-term annealing of these specimens has little affect on the interfacial microstructure other than to cause the reaction zone to thicken slightly. Upon oxidation of the couple, a NiO sub-layer forms just beneath the braze-5YSZ reaction zone. The formation of this oxide appears to be quite aggressive, penetrating along grain boundaries into the braze filler material and consuming a substantial amount of nickel and titanium from the underlying filler material.

(3) Joining of the braze to stainless steel gives rise to substantial iron and chromium alloying in the braze due to partial dissolution of the metal substrate. Consequently, the oxidation behavior of the braze is substantially changed. Upon high temperature air exposure, a protective Cr_2O_3 scale grows on the free surface of the braze-side of the couple.

(4) The presence of the stainless steel in the Au ABA joining specimens dramatically alters the microstructures and compositions of the braze at the 5YSZ/braze interface. In the annealed specimens, the TiO_2 interfacial layer previously found in the 5YSZ wetting specimens does not exist. Instead, two phases are observed at the interface, a mixed iron-nickel oxide precipitate in a gold-based matrix; both containing measurable amounts of chromium. The titanium appears to be tied up in two separate phases not found directly at the interface with the 5YSZ, microscale TiO_2 and Widmanstätten-type iron-titanium precipitates. Upon exposure to air at 800°C, a chromia layer develops along the former 5YSZ faying surface and appears to protect the rest of the brazed joint from the type of aggressive oxidation seen in the Au ABA/5YSZ coupons.

ACKNOWLEDGMENTS

The authors would like to thank Jim Coleman, Nat Saenz, and Shelly Carlson for their assistance in metallographic and SEM sample preparation and analysis, Kerry Meinhardt for his assistance in fabricating the YSZ discs, and Dean Paxton for his help in fabricating the braze specimens. This work was supported by the U.S. Department of Energy, Office of Fossil Energy, Advanced Research and Technology Development Program. The Pacific Northwest National Laboratory is operated by Battelle Memorial Institute for the United States Department of Energy under Contract DE-AC06-76RLO 1830.

REFERENCES

1. C. W. Fox and G. M. Slaughter, *Weld. J.*, 43, 591 (1964).
2. O. M. Akselsen, *J. Mater. Sci.*, 27, 1989 (1992).
3. K. D. Meinhardt, J. D. Vienna, T. R. Armstrong, and L. R. Peterson, PCT Int. Appl., WO01909059 (2001).
4. R. R. Kapoor and T.W. Eagar, *J. Am. Ceamr. Soc.*, 72, 448 (1989)
5. A. J. Moorehead and H. Kim, *J. of Mater. Sci.*, 26, 4067 (1991)
6. D. B. Lee, J.H. Woo, and S.W. Park, *Mater. Sci. Eng.*, A268, 202 (1999)
7. O. Kubashewski and B. E. Hopkins, *Oxidation of Metals and Alloys, 2nd ed.,* Butterworth, London, (1962)

TiO_2-MODIFIED Ag-CuO REACTIVE AIR BRAZES FOR IMPROVED WETTABILITY ON MIXED IONIC/ELECTRONIC CONDUCTORS

J.S. Hardy, K. S. Weil, and J. Y. Kim, E.C. Thomsen
Pacific Northwest National Laboratory
902 Battelle Blvd.
Richland, WA 99352

J. T. Darsell
Washington State University
PO Box 642920
Pullman, WA 99164-2920

ABSTRACT

Mixed ionic/electronic conducting perovskite oxides such as lanthanum strontium cobalt ferrite (LSCF) and doped barium cerates are candidate functional materials for electrochemical devices such as gas separation membranes and solid oxide fuel cells (SOFCs). However, taking full advantage of the unique properties of advanced ceramics such as mixed conducting oxides depends in large part on developing reliable joining techniques that can be fired and operated under high temperature exposure to air. An electrically conductive ceramic-to-metal joining material would further benefit the device because it would allow current to be either drawn from or carried to the functional mixed conducting oxide component in these devices without requiring an additional current collector. Reactive air brazing (RAB) is a joining technique that can be carried out in air using AgCu-based brazing alloys that wet oxide ceramic surfaces, creating a joint with a non-brittle filler material that is electrically conductive. The effect of adding 0.5 mol% Ti to the RAB alloy on surface wettability, long-term electrical resistivity at 750°C, and reactivity with the $La_{0.6}Sr_{0.4}Co_{0.2}Fe_{0.8}O_{3-\delta}$ (LSCF-6428 or LSCF) substrates will be discussed.

INTRODUCTION

Because of their unique electrochemical properties, the mixed ionic/electronic conductor (MIEC) family of materials, which includes complex metal oxide perovskites such as LSCF-6428 ($La_{0.6}Sr_{0.4}Co_{0.2}Fe_{0.8}O_{3-\delta}$), represents a large potential market. The estimated demand for MIEC oxide-based sensors has grown to approximately \$3 billion dollars[1]. SOFCs could represent an even larger market potential with MIEC oxides under consideration as cathode materials[2] and as agents to catalyze fuel reformation at the SOFC electrodes. Moreover, electrically driven oxygen-ion transport membranes for oxygen gas separation, partial hydrocarbon oxidation, and waste reduction and recovery can employ MIEC oxides[3]. MIEC oxide-based membrane technology offers the potential to separate oxygen from air with far greater efficiency and at one-third lower cost than the cryogenic processing technology used today. And unlike cryo-separation, oxygen transport membranes operate at high temperature, making them ideally suited for direct integration with coal gasification plants[4].

Reliable joing methods are required if such devices are to exploit the unique properties of these advanced ceramics. However, high operation temperatures and low tolerances of many MIEC compositions for reducing environments severely limit the number of joining technologies that can be used. For many MIEC oxides, joining must take place in air to avoid reduction, phase separation, and consequential degradation of the ceramic. Furthermore, in evaluating candidate joining techniques, one must remember that the joint must withstand stresses imposed by thermal cycling and, in many cases, must also hermetically seal the device to maintain the appropriate chemical gradient, typically oxygen, across the electrochemically active membrane enabling the device's operation. In addition, when used as a current collector or an electrical connection between MIEC components, the joint should exhibit long-term electrical conduction with minimal ohmic loss under a moderate-to-high current density while remaining chemically stable at the high temperature, oxidizing operating conditions of the device.

Reactive air brazing (RAB) forms a predominantly metallic joint directly in air without the need for fluxing[5]. The process utilizes braze compositions that, in the molten state, consist of a metal oxide that is at least partially dissolved in a noble metal filler material. Ag-CuO has been found to be one such material system that is well suited for the RAB process. This study demonstrates that the addition of a small amount (0.5 mol%) of Ti to the Ag-Cu alloy further enhances the surface wetting of the braze on LSCF. Like Cu, Ti also oxidizes in-situ during firing to form its oxide. Schueler et al.[6] were the first to report using Ag-CuO to join ceramics in air, demonstrating this capability on Al_2O_3 substrates. More recently, Erskine et al.[7] used Ag-CuO to braze an electrostrictive ceramic composition having the perovskite crystal structure that is also characteristic of many MIEC oxides, including LSCF-6428.

EXPERIMENTAL

LSCF (99.9% purity; Praxair Specialty Ceramics, Inc.) pellets were pressed uniaxially at 25 MPa, then isostatically at 135 MPa. After sintering at 1300°C for 4 h, the pellets were polished on one face to a >3 μm finish. Reactive air braze pellets for sessile drop wetting experiments were pressed from mixtures of copper (99% purity; 1 – 1.5 μm average particle size; Alfa Aesar), silver (99.9% purity; 0.5 – 1 μm average particle size; Alfa Aesar), and TiH_2 (98% Purity; ~325 mesh; Aldrich) powders that had been ball-milled in the desired molar ratios in ethanol for 24 h, dried, and sieved through a 200 mesh screen. The braze compositions evaluated in this study adhere to the general chemical formula, $(Ag_{1-x}Cu_x)_{0.995}Ti_{0.005}$. The metallic copper readily oxidizes to form CuO during air firing while the TiH_2 oxides to form TiO_2. These metal oxide constituents are then present in the same molar ratio to Ag as were the source powders. The braze powder mixtures were pressed into 7 mm diameter by 7 mm thick pellets for sessile drop testing or mixed with a polymer binder (B75717, Ferro Corp.) in the proper ratios to produce a 70 wt% solids paste for brazing electrical test samples.

Sessile drop wetting experiments were performed by placing RAB pellets of varying Cu content upon the center of the polished face of a LSCF pellet measuring approximately 1" in diameter. The melting and wetting behavior of the braze pellet were recorded using a high speed video camera that was aligned to view the sample through a quartz window in the door of a static air box furnace. Samples were heated to 900°C at 25°C/min at which point the heating rate was slowed to 10°C/min. At 1000°C, 1050°C, and 1100°C, the temperature was held for 10 min to allow the molten braze bead to reach its equilibrium shape. Frames from the final seconds of each 10 min dwell were captured from the recorded video and converted to computer images using VideoStudio6™ (Ulead Systems, Inc.) video editing software. These images were imported into Canvas™ (version 9.0.1, build 689; Deneba Systems, Inc.) graphics software which provided tools that facilitated contact angle measurements.

Four probe resistance measurements were used to assess the effects of long term exposure to electrical current at a typical operating temperature (750°C) on the electrical properties of Ti-modified RAB joints to LSCF. Samples for electrical testing were prepared by brazing the polished faces of LSCF pellets together at 1050°C for 30 min in air using braze paste. A 1.5 mm dia. hole had been drilled into the center of the opposite, unpolished face of each pellet to a depth that was 1 mm less than the thickness of the pellet. Platinum paste was applied to the top and bottom of the LSCF/RAB/LSCF sandwich samples and fired on at 900°C for 30 min to act as contact electrodes. A Pt foil with a hole cut out of the center to match the one drilled in the brazed samples made contact to the fired Pt paste electrodes on the faces of the sample. A platinum probe wire that had been welded to form a small bead of platinum at the end was inserted, bead first, down to the bottom of the drilled hole on each side of the sample, thereby functioning as a reference electrode. Electrical leads attached to the Pt foil contacts and the Pt probe reference electrodes were connected to a potentiostat (Arbin Instruments, Model BT2043) which applied an electrical current through the Pt foil contacts and measured the voltage through the Pt probes, thereby establishing a four-probe resistance measurement configuration as depicted in fig. 1. A constant current of 1 A was applied to the sample for 250 h while the sample was held under isothermal conditions at 750°C. Voltage measurements were taken at periodic intervals throughout the 250 h test.

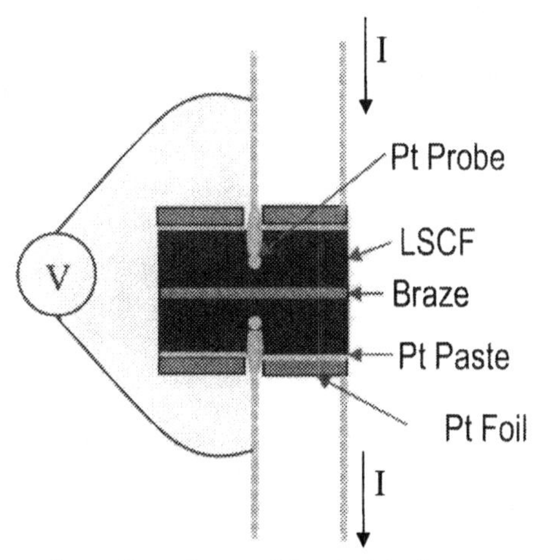

Figure 1 The sample configuration used for electrical measurements.

A scanning electron microscope (JEOL JSM-5900LV) equipped with an Oxford energy dispersive X-ray analysis (EDX) system, which employs a windowless detector for quantitative detection of both light and heavy elements, was employed to examine the microstructure and phase composition at the interfaces between LSCF substrates and reactive air brazes.

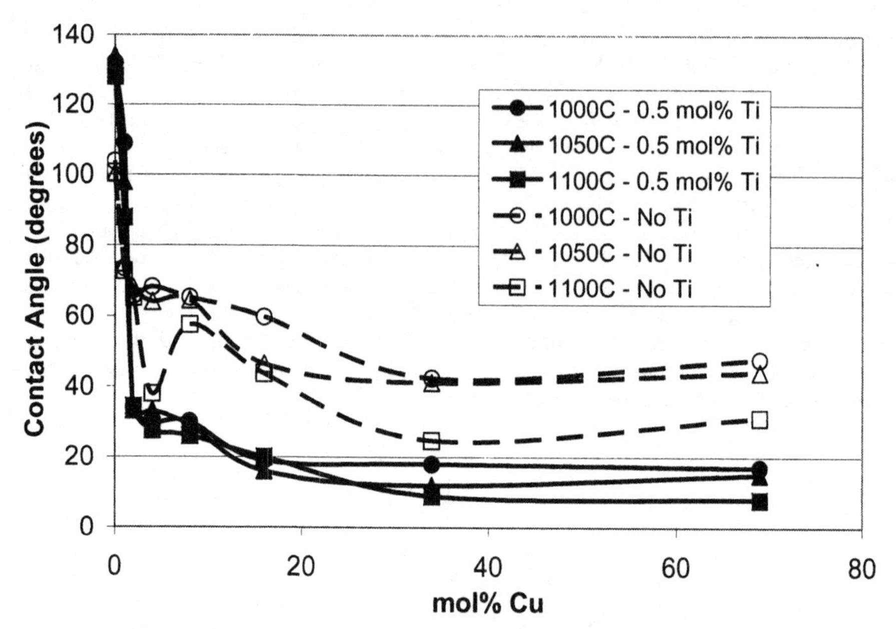

Figure 2 The effect of CuO content on the contact angle of Ag-based RAB compositions containing 0.5 mol% Ti (solid lines and symbols) and no Ti (dashed lines and hollow symbols) on the surface of LSCF.

RESULTS AND DISCUSSION

Figure 2 shows the measured contact angles of molten RAB compositions on polished LSCF surfaces as a function of Cu-content. The solid lines and symbols represent the brazes that contained 0.5 mol% TiO_2, while the dotted lines and hollow symbols represent samples with no TiO_2. The TiO_2-modified brazes containing no Cu were clearly non-wetting on the surface of LSCF at any temperature measured, having a contact angle of around 130° which is well above the generally accepted 90° threshold of wettability. With no Cu added, the 0.5 mol% TiO_2 addition actually caused an increase in the contact angle of silver on LSCF. With the addition of only 1 mol% CuO to the TiO_2-modified RAB, a dramatic decrease in contact angle is measured, although it still straddles the wettability threshold of 90°. At 1 mol% CuO, samples with no TiO_2 still exhibited a contact angle that at was at least 15° lower than that of the TiO_2-modified braze indicating that TiO_2-additions still detracted from wettability in 1 mol% CuO braze. Not until 2 mol% CuO does the TiO_2 braze show definite surface wetting on LSCF and begin to promote a dramatic improvement in wetting over RAB containing no TiO_2. The contact angles measured for the 2 mol% Cu sample containing TiO_2 are over 30° lower than those of the corresponding sample with no TiO_2. Above 2 mol%, the contact angle continues to decline with increasing CuO, albeit at a slower rate. A definite elbow in the curve exists between 2 and 4 mol% CuO where the effect of CuO-content on contact angle is no longer as pronounced as it was at lower CuO concentrations. For RAB compositions with Cu concentrations of at least 2 mol%, there is a distinct improvement in the wetting of LSCF for brazes containing 0.5 mol% TiO_2. Temperature did not seem to have any definite, consistent effect on contact angle and all compositions achieved their equilibrium shape almost immediately upon reaching target temperatures. This suggests that the surface modification that results from reaction with the oxide components of the RAB alloys probably occurs soon after the braze is molten and the oxides become free to migrate to the interface. The fact that the addition of TiO_2 did not improve the wettability of the RAB compositions containing less than 2 mol% CuO, and in fact made it worse, suggests that an interaction between the TiO_2 and CuO brings about the drastic improvement in wettability and not simply TiO_2 acting independently.

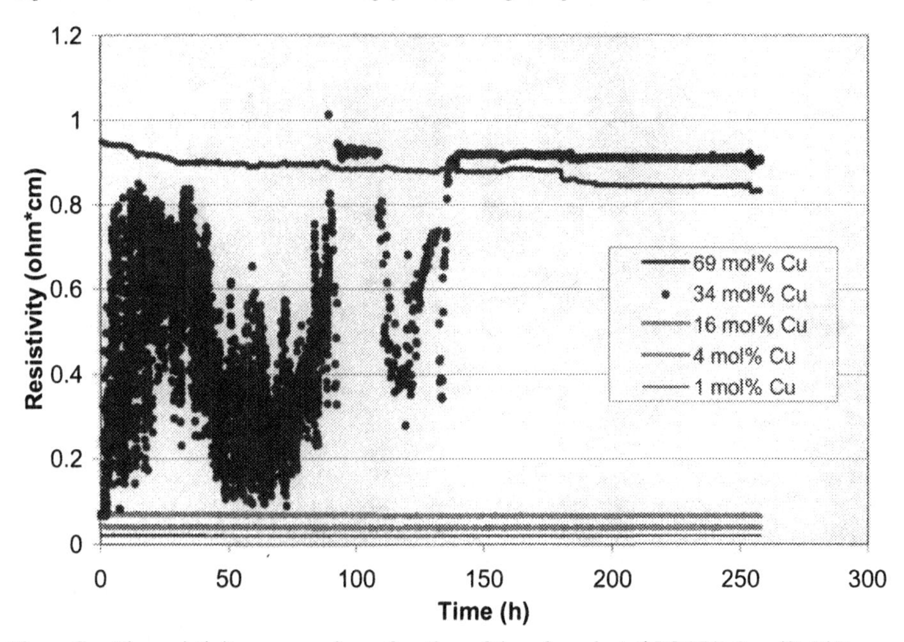

Figure 3 The resistivity measured as a function of time for selected LSCF/0.5 mol% TiO_2-modified RAB/LSCF composite samples.

The plots in fig. 3 show how resistivity changes over long term exposure to a current of 1 A and a typical MIEC oxide device operation temperature of 750°C. The measurements shown in the plot represent resistivity values for the layer of braze with 1 mm of LSCF on either side, inclusive. By including the thin layer of LSCF on either side of the braze, any chemical or physical changes in the LSCF that might manifest themselves through changes in the resistivity are also detected. It was found that the resistivity in samples with up to 16 mol% Cu were largely unaffected over time during long term operation at 750°C and increased in magnitude with increasing CuO-content. Increased resistivity with more CuO was expected because CuO is far less conductive than either the filler metal or the LSCF. The sample with 34 mol% Cu showed wide variations in resistivity over the first approximately 140 h of testing. However, for the last 100 h of testing it settled into a stable value of slightly greater than 0.9 ohm*cm. The observed fluctations may have been due to poor electrical contact or possibly to reaction between the CuO and the LSCF pellets. The sample with 69 mol% Cu had an initial resistivity of about 0.95 ohm*cm which, over the 250 h test period, gradually declined to about 0.85 ohm*cm. The significantly higher resistivity values of samples with 34 and 69 mol% Cu are due to the formation of a continuous CuO phase in these samples which impedes electrical conductivity.

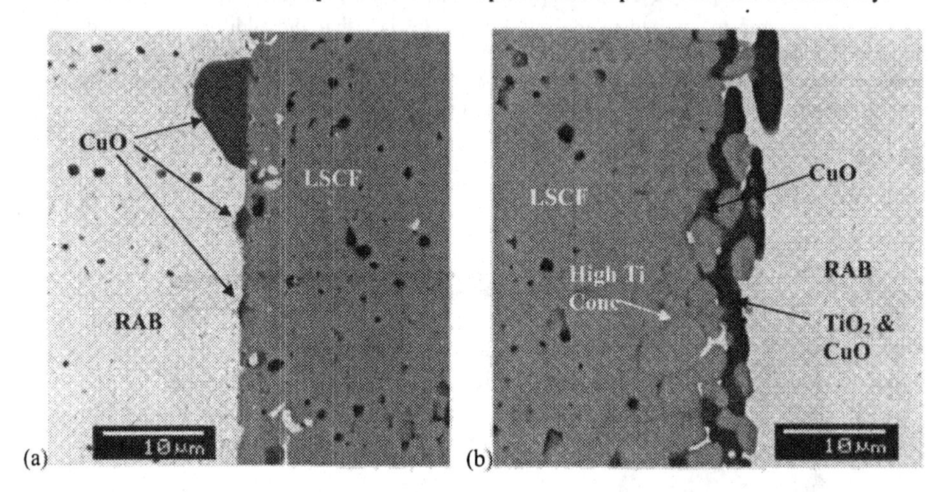

Figure 4 SEM images of samples of LSCF after brazing with 4 mol% CuO RAB compositions. (a) no TiO_2, (b) 0.5 mol% TiO_2.

Figure 4 compares the microstructures of 4 mol% CuO brazes with or without TiO_2-additions. One of the more noteworthy differences between the two samples is the presence of dark lines outlining the grains in the sample that was brazed with a TiO_2-modified braze. It was determined by EDS compositional analysis that the dark areas correspond to a high concentration of Ti, suggesting that the Ti from the braze diffuses along the grain boundaries of the LSCF. The presence of LSCF grains that have apparently been dislodged from the main body of LSCF in the TiO_2 sample suggest that the diffusing Ti may weaken the grain boundaries. A similar phenomenon is not observed in the sample with no TiO_2. In the TiO_2 sample, it was also observed that the oxide coverage of the LSCF surface appears to be much more continuous than in the sample with no TiO_2.

CONCLUSIONS

The addition of as little as 0.5 mol% Ti to a Ag-Cu braze alloy greatly improves surface wetting on LSCF when the braze contains at least 2 mol% Cu. In samples with 0 and 1 mol% Cu, the Ti-addition was actually deleterious to wetting, suggesting that it is an interaction between TiO_2 and CuO that results in improved wetting and not the TiO_2 acting independently. Resistivity was found to increase with increasing CuO-content in TiO_2 modified brazes. In samples with up to 16 mol% CuO, the resistivity was relatively low and stable for 250 h at 750°C under a constant current of 1 A. Microstructural analysis revealed that the Ti from the braze diffuses along the grain boundaries of LSCF, possibly weakening them mechanically.

ACKNOWLEDGMENTS

The authors would like to thank Nat Saenz, Shelly Carlson, and Jim Coleman for their assistance in sectioning and polishing the resistance samples and conducting the metallographic and SEM analysis work. This work was supported by the U. S. Department of Energy, Office of Fossil Energy, Advanced Research and Technology Development Program. The Pacific Northwest National Laboratory is operated by Battelle Memorial Institute for the United States Department of Energy (U.S. DOE) under Contract DE-AC06-76RLO 1830.

REFERENCES

[1]G.R. Doughty and H. Hind, "The Applications of Ion-Conducting Ceramics," *Key Engineering Materials*, **122-124** 145-62 (1996).

[2]N.Q. Minh, "Ceramic Fuel Cells," *Journal of the American Ceramic Society*, **76** 563-88 (1993).

[3]A.V. Kovalevsky, V.V. Kharton, V.N. Tikhonovich, E.N. Naumovich, A.A. Tonoyan, O.P. Reut, and L.S. Boginsky, *Materials Science and Engineering*, **B52** 105-6 (1998).

[4]A.K. Anand, C.S. Cook, J.C. Corman, and A.R. Smith, *Transactions of the ASM. Journal of Engineering for Gas Turbines and Power*, **118** 732-6 (1996).

[5]K. S. Weil and D. M. Paxton, *Proceedings of the 26th Annual Conference on Composites, Advanced Ceramics, Materials, and Structures: A*, American Ceramic Society, Westerville, OH (2002).

[6] C.C. Shüler, A. Stuck, N. Beck, H. Keser, and U. Täck, "Direct Silver Bonding – An Alternative for Substrates in Power Semiconductor Packaging," *Journal of Materials Science: Materials in Electronics*, **11** [3] 389-96 (2000).

[7] K.M. Erskine, A.M. Meier, and S.M. Pilgrim, "Brazing Perovskite Ceramics with Silver / Copper Oxide Braze Alloys," *J. Mater. Res.*, **37** 1705-1709 (2002).

MICROSTRUCTURE, MELTING AND WETTING PROPERTIES OF Pd-Ag-CuO AIR BRAZE ON ALUMINA

Jens T. Darsell
Washington State University
PO Box 642920
Pullman, WA 99164-2920

John S. Hardy, Jin Y. Kim, and K. Scott Weil
Pacific Northwest National Laboratory
P.O. Box 999
Richland, WA 99352

ABSTRACT

A new ceramic brazing technique referred to as reactive air brazing (RAB) has recently been developed for potential applications in high temperature devices such as gas concentrators, solid oxide fuel cells, gas turbines, and combustion engines. At present, the technique utilizing a silver-copper oxide system is of great interest. The maximum operating temperature of this system is limited by its eutectic temperature of ~935°C, although in practice the operating temperature will need to be lower. An obvious strategy that can be employed to increase the maximum operating temperature of the braze material is to add a higher melting noble alloying element. In this paper, we report the effects of palladium addition on the melting characteristics of the Ag-CuO system and on the wetting properties of the resulting braze with respect to alumina. It was found that the addition of Pd will cause an increase in the melting temperature of the Ag-CuO braze but possibly at a sacrifice of wetting properties depending on composition.

INTRODUCTION

As the operating temperature of high temperature equipment, such as molten metal gas sensors, gas turbines, internal combustion engines, and heat exchangers, continues to be pushed upward by efficiency considerations, there is an ever-increasing need to incorporate ceramic materials into device components. Ceramics are attractive materials for these applications because of their excellent high-temperature mechanical properties and their high level of wear and corrosion resistance. Their usefulness, however, is limited by the current inability to economically manufacture large or complex-shaped ceramic components that exhibit reliable performance. One alternative is to fabricate small, simple-shaped parts that can be assembled and joined to form a larger, more complex structure. Conventional fusion welding is not feasible since the high melting point of the ceramic would impose extremely high power requirements and the chemical and mechanical nature of the metal and ceramic constituents is not simultaneously compatible with this joining technique. Diffusion bonding of ceramics to metals is possible in some cases, although the durability of the bond at high temperature, particularly under oxidizing conditions, is questionable. In addition, the equipment costs associated with this process are high due to the stringent operating conditions which include high temperature and high pressure in a vacuum environment.

One of the most reliable and best understood methods of joining dissimilar materials is brazing. In this technique, a filler metal, whose liquidus is well below that of the materials to be joined, is molten so that the braze is allowed to flow and fill the gap between the two faying surfaces under capillary action, and is finally cooled to solidify the joint. Unfortunately, when applied to the bonding of ceramics, brazing has historically suffered a serious drawback in that most commercial brazes do not wet ceramic surfaces. Over the past several years, considerable efforts have been focused on overcoming this issue. It has been found that the addition of a reactive metal, such as Ti or Zr (known as active metal brazing), results in the reduction of the ceramic phase at the joining interface to create an intermediate layer that is in chemical equilibrium with both the ceramic and braze metal phases. Once this reduced phase forms, wetting of the ceramic surface by the filler metal tends to be greatly improved.

While active metal brazing provides a solution to the problem of wettability, it presents a serious drawback in terms of processing: atmospheric control is required to protect this intermediate joining layer from oxidation at operating temperature. Recently, a new brazing technique for joining ceramics, referred to as reactive air brazing (RAB) was developed, inherently exhibiting resistance against degradation due oxidation at high temperature. RAB consists of two components, an inert noble metal filler material such as Ag, Au, or Pt, and an element prone to oxidation such as Cu, the oxide of which reactively modifies ceramic faying surfaces. Brazing is therefore conducted directly in air without need for atmospheric control or surface cleaning fluxes. During the joining process, the alloying element oxidizes in-situ to form a reactive oxide that modifies the ceramic surface and improves the wetting characteristics of the braze.

One particular system of interest in the development of RAB is the binary Ag-CuO system. Meier et al have shown that increasing the CuO content significantly improves the wetting behavior of Ag-CuO on alumina[1] in an inert environment. The Ag-CuO system was demonstrated to be capable of joining alumina in air[2, 3, 4] and to have improvements in wetting behavior with CuO additions[4]. However, the applicability of the Ag-CuO braze is limited by its eutectic temperature that occurs at ~935 °C. To extend the use of this braze to higher temperatures, it is necessary to increase the melting point of the braze. This can be achieved by addition of an element such as palladium, which has a higher melting temperature. In this study, the effects of palladium addition on the liquidus and subsequent wetting characteristics of the resulting braze were investigated using alumina as a model oxide substrate.

EXPERIMENTAL

Materials

Sessile drop experiments were performed on polycrystalline alumina discs (Al-23, Alpha Aesar, Ward Hill, MA 01835, 98% dense, 99.7 % pure) that contained a small amount of silicate as an impurity. These discs were 50 mm in diameter and 6 mm thick. One surface of the disc, on which a braze pellet was placed for the contact angle measurement, was polished to a 1 μm finish using water-based diamond suspensions. The discs were then cleaned with acetone, and rinsed with propanol, air dried, and finally heated in static air to 600 °C for 4 hours to burn off any residual organic species. Alumina crucibles (Netzsch, Burlington MA) used in the differential scanning calorimetry (DSC) experiments were also preheated to remove organic contamination.

As listed in Table I, various braze compositions were selected. These compositions were formulated by dry mixing the appropriate amounts of silver powder (99.9%, 0.75 μm average particle size, Alpha Aesar), copper powder (99%, 1.25 μm average particle size, Alpha Aesar) and palladium powder (submicron, 99.9+%, Aldrich). These mixed powders were then cold-pressed into pellets measuring approximately 7 mm in diameter and 10 mm tall for sessile drop experiments. For DSC experiments, 2 mm diameter pellets were prepared from approximately 5-10 mg of braze powder in the same manner.

Table I. Braze compositions employed in this study

Braze I.D.	Ag Content (mol %)	Pd Content (mol %)	CuO Content (mol %)
Ag-1Cu	99	0	1
1Pd:3Ag-1Cu	74.2	24.8	1
1Pd:1Ag-1Cu	49.4	49.6	1
Ag-1.4Cu	98.6	0	1.4
1Pd:3Ag-1.4Cu	73.9	24.7	1.4
1Pd:1Ag-1.4Cu	49.2	49.4	1.4
Ag-2Cu	98	0	2
1Pd:3Ag-2Cu	73.4	24.6	2
1Pd:1Ag-2Cu	48.9	49.1	2
Ag-4Cu	96	0	4
1Pd:3Ag-4Cu	71.9	24.1	4
1Pd:1Ag-4Cu	47.9	48.1	4
Ag-8Cu	92	0	8
1Pd:3Ag-8Cu	68.9	23.1	8
1Pd:1Ag-8Cu	45.9	46.1	8
90Ag10Cu	90	0	10
1Pd:3Ag-10Cu	67.4	22.6	10
1Pd:1Ag-10Cu	44.9	45.1	10
Ag-20Cu	80	0	20
1Pd:1Ag-20Cu	39.9	40.1	20
Ag-30Cu	70	0	30
1Pd:3Ag-34Cu	49.4	16.5	34.1
1Pd:1Ag-34Cu	32.9	33.1	34
Ag-40Cu	60	0	40
Ag-50Cu	50	0	50
Ag-60Cu	40	0	60
Ag-69Cu	30.65	0	69.35
Ag-80Cu	20	0	80

Characterization

DSC analysis was conducted using a Netzsch system (model STA 449C Jupiter) equipped with a high temperature furnace and a Type-S sample carrier. All the experiments were performed in flowing dry air at a flow rate of 10 ml/min. A heating rate of 10 °C/min was employed in each run, and the maximum temperature for the experiment was determined based on the Pd content of the sample.

Sessile drop experiments were conducted in a static air muffle furnace, furnished with a quartz window through which the contact angle of heated specimen could be observed. The braze pellets were placed on the polished face of the substrate and heated in the furnace. The heating cycle employed for each sample was dependent on the braze composition. For example in samples containing no palladium, the furnace was heated at 30 °C/min to an initial temperature of 900 °C and held for 15 min then subsequently heated at 10 °C/min to soak temperatures of 950 °C, 1000 °C, 1050 °C, and 1100 °C, at each of which the furnace was held for a period of 15 min. For samples containing 25 mol % palladium in silver, the furnace was heated at 30 °C/min to an initial temperature of 1100 °C, followed by a heating at 10 °C/min, with 15 min soaks at 1150 °C, 1200 °C, and 1250 °C. For samples containing 50 mol % Pd in Ag, the same heating cycle was used with the addition of 15 min holds at 1300 °C, and 1350 °C. A video camera with a zoom lens recorded the profile of the specimens during the heating cycle. Selected frames from the videotape were converted using Ulead™ software to digital images from which the contact angle between the braze and alumina substrate could be measured.

Microstructural analysis was performed on polished cross sections of the wetting samples, using a scanning electron microscope (SEM, JEOL, JSM-5900 LV) equipped with an energy dispersive X-ray (EDX) detector and analysis system. To avoid electrical charging of the samples in the SEM, they were carbon coated and grounded.

RESULTS AND DISCUSSION

Melting Behavior

Figure 1 shows DSC curves, exhibiting endothermic peaks corresponding to melting of pure Ag, Ag with 1 mol % CuO, the 1:3 Pd:Ag alloy with 1 mol % CuO, and the 1:1 Pd:Ag alloy with 1 mol % CuO. To obtain the onset of melting, the first derivative of each DSC curve was calculated by differentiating the DSC curve with respect to temperature. The onset was identified as a temperature at which the derivative of the DSC curve revealed deviation from the baseline. The liquidus temperature was defined by the temperature at which the endothermic peak re-joins the baseline. Some samples such as the Ag-1 mol % CuO exhibited an endotherm with double peaks. For these samples, the solidus temperature was defined as the temperature where the derivative of the first peak deviated from baseline, while the liquidus temperature was defined as the point at which the derivative of the second peak returned to baseline.

Solidus and liquidus temperatures of Pd-Ag compositions obtained in this study are plotted as a function of Ag content in the range of 50 ~ 100 mol %, and also compared with values reported by Karayaka and Thompson[5] in Figure 2. The solidus temperature obtained for pure Ag was 950 °C, which agrees well with values regularly obtained on the Netzsch instrument of 951 °C. A liquidus temperature for pure Ag was found to be 961 °C which agrees with literature values[6].

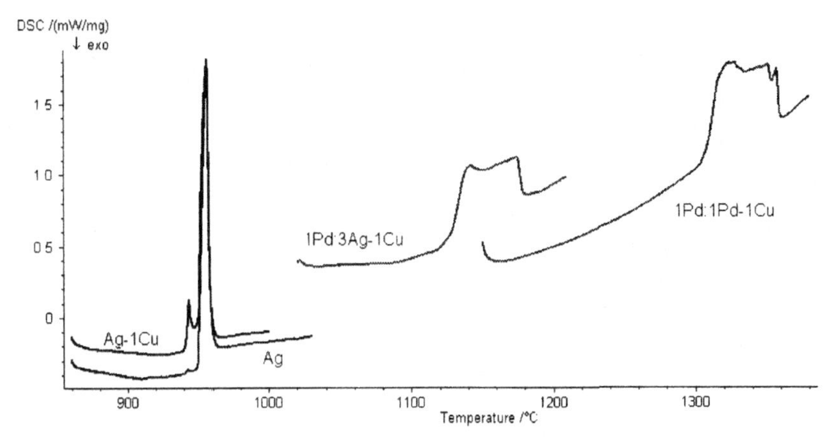

Figure 1 DSC curves obtained from pure Ag, Ag with 1 mole % CuO, the 1:3 Pd:Ag alloy with 1 mol% CuO, and the 1:1 Pd:Ag alloy with 1 mole % CuO.

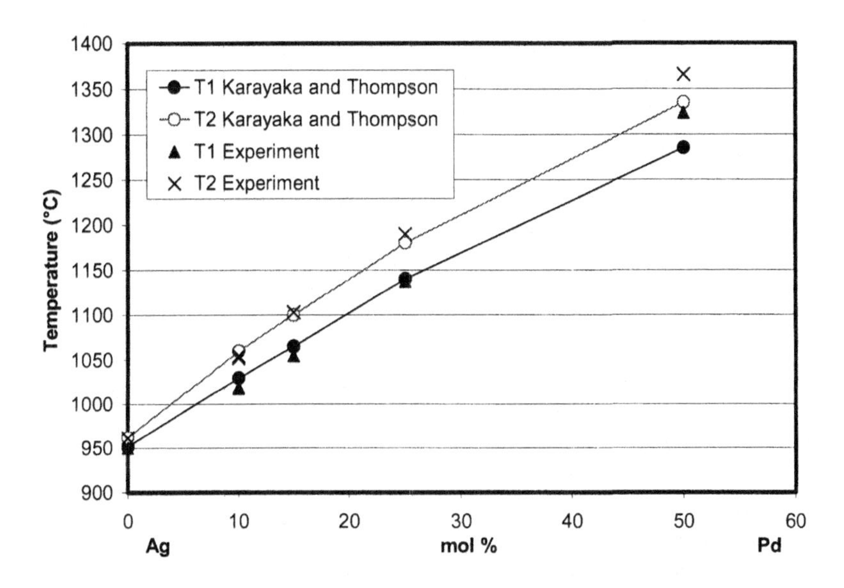

Figure 2 Solidus (T1) and liquidus (T2) temperatures of Ag-Pd obtained in the DSC analysis. For comparison, values reported by Karyaka and Thomson[5] are also plotted along with our experimental values.

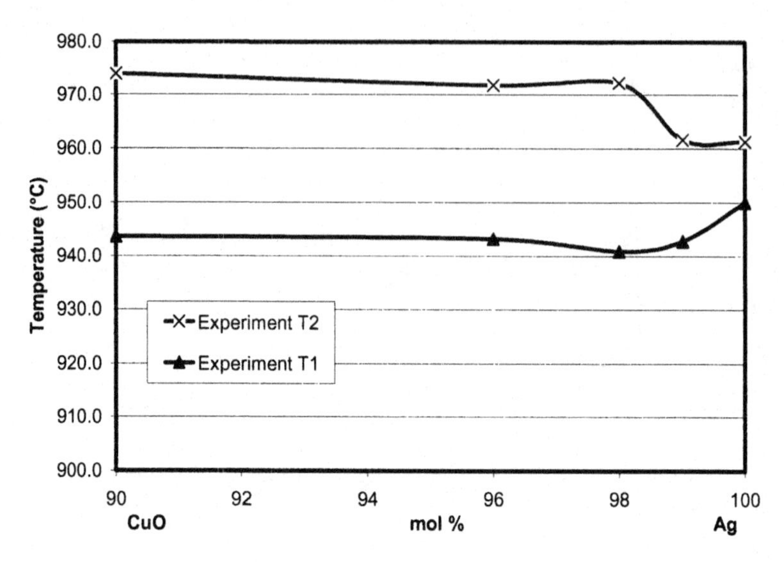

Figure 3 Solidus (T1) and Liquidus (T2) temperatures of the Ag-CuO system obtained in the DSC analysis.

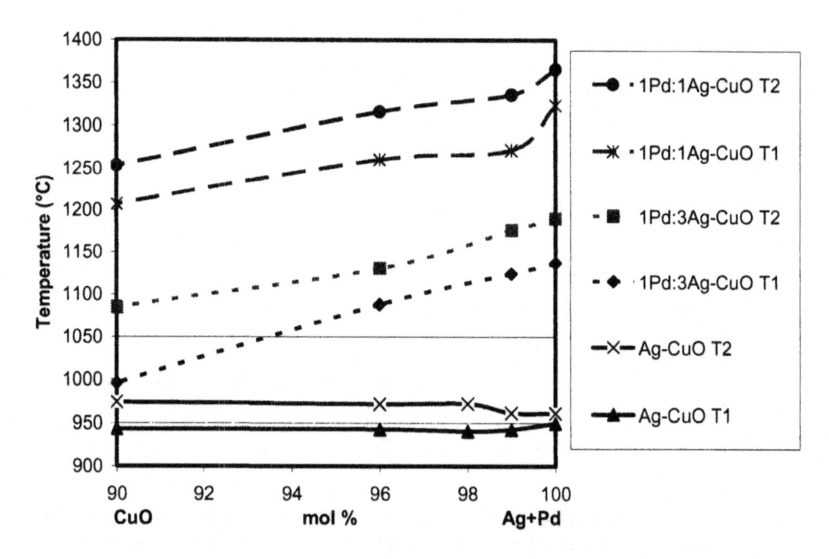

Figure 4 Experimental solidus (T1) and Liquidus (T2) temperatures for Pd-Ag-CuO containing varying metal (Ag + Pd) compositions and Pd:Ag ratios.

Shown in Figure 3 are the solidus (T1) and liquidus (T2) temperatures obtained for the Ag-CuO samples containing 90, 96, 98, and 100 mol % Ag. The figure indicates that the onset of melting occurs at approximately 945 ± 5 °C and the liquidus temperature is approximately 972 ± 2 °C for the samples with Ag contents of 90-98 mol %, while the liquidus temperature for the 99 mol % Ag sample is reduced to 961 °C. This result agrees well with the data reported by Nishiura et al.[7]. This result validates the accuracy of our method for obtaining the solidus and liquidus temperatures in this study.

The solidus and liquidus temperatures of the two series of Pd-Ag-CuO compositions are shown as a function of noble metal content (Ag + Pd) in Figure 4. A significant increase in the solidus and liquidus temperatures is observed as the Pd to Ag ratio increases. The increase is less dramatic at higher CuO contents (low metal contents).

Wetting Properties

Results from the sessile drop experiments are plotted as a function of metal contents in Figure 5. An increase in contact angle is observed with increasing Pd content in the braze. All of the samples with a Pd:Ag ratio of 1:3 revealed a wetting condition (contact angle < 90°), although the contact angle is approximately 5-20° larger than the comparable Ag-CuO braze in the metal content between 66 and 96 mol %. In the case of samples containing 98 and 99 mol % metals, the contact angle obtained from the Pd:Ag specimens with a 1:3 ratio is almost identical to that of the Ag-CuO system. On the other hand, the samples with a higher Pd content (Pd:Ag = 1:1) exhibited a significant increase in the contact angle, which was on average 10-60°larger than that of the Ag-CuO system. This effect was especially severe at low CuO contents, displaying non-wetting (contact angle > 90°) when the CuO content is less than 10 mol %.

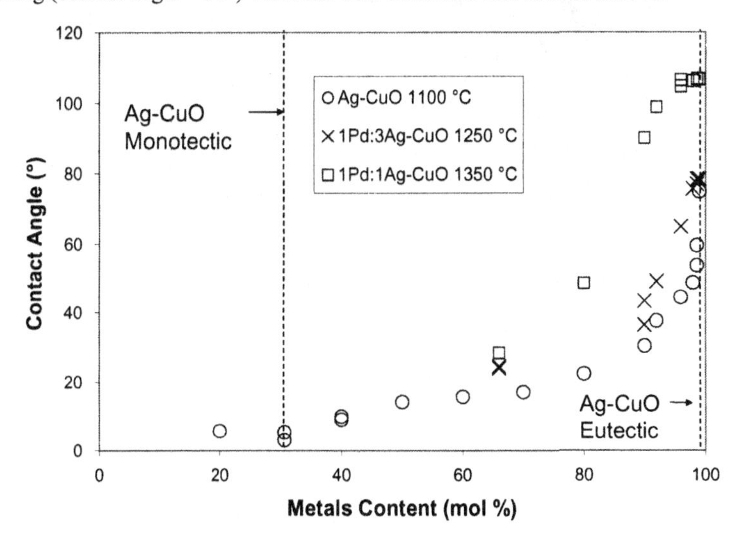

Figure 5 Contact angle of the Pd-Ag-CuO braze with various Pd/Ag ratios as a function of total metal contents. Samples held at final temperatures (1100 °C for Ag-CuO, 1250 °C for 1Pd:3Ag-CuO and 1350 °C for 1Pd:3Ag-CuO) for 15 min.

Microstructure

Figure 6 shows SEM images collected on the braze/alumina interfaces with brazes containing 4 mol % CuO. The Ag-4CuO sample exhibits no porosity at the interface indicating good bonding between the braze and substrate as see in Figure 6(a). The CuO phase in the braze exists in two forms; precipitates along the alumina-braze interface and inclusions in the Ag phase. A trace amount of copper was also observed within the alumina substrate along the grain boundaries, indicating possible migration of the CuO along the grain boundaries of alumina.

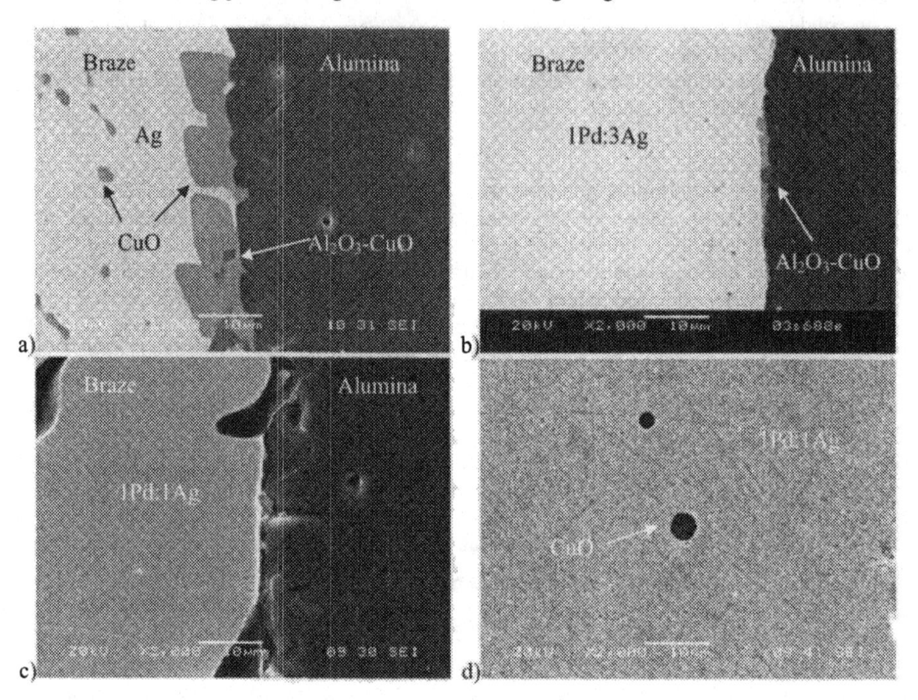

Figure 6 Scanning electron micrographs at 2000X of the braze/alumina interface of sessile drop samples containing 4 mol % of CuO and a) Ag, b) 1Pd:3Ag, and c) 1Pd:1Ag as well as d) the precipitates containing Cu in the 1Pd:1Ag-Cu sample.

The 1Pd:3Ag sample also exhibits excellent bonding of the braze to the substrate as indicated by the lack of porosity (refer to Figure 6(b)). EDX analysis shows a metal matrix containing approximately 25 mol % Pd and 70 mol % Ag. There are also inclusions in the metal matrix, which contain Cu and O in a ratio of approximately 2 to 1 possibly indicating the presence of Cu_2O. In addition, two different interfacial products were observed: (1) a lighter gray layer containing Cu, Al, and O and (2) darker precipitates also containing Cu, Al, and O but with a larger amount of O and the addition of a trace of Mg.

On the other hand, the 1Pd:1Ag-4CuO sample shows poor bonding between the braze and substrate (Figure 6(c)). This result was expected since this sample exhibited poor wettability in comparison with the other brazes with no Pd or a lower Pd content (refer to Figure 5). EDX analysis shows that the braze matrix (a light gray region) contains Ag and Pd in a ratio of 1 Pd to 1 Ag. The particles at the interface between the braze and alumina contain mostly Al and O with a trace of Si, but no Cu is observed at this site. Cu was mostly observed in a form of precipitates within the braze matrix as can be seen in Figure 6(d). These precipitates contain Cu and O, and a trace of Si.

CONCLUSIONS

This addition of palladium significantly influenced melting behavior, wetting properties and microstructure of the braze. A significant increase in the solidus and liquidus temperatures was observed as the Pd to Ag ratio increased but this increase was less dramatic at higher CuO contents (low metal contents). In the case of samples containing a Pd:Ag ratio of 1:1, the contact angle was 10-60°larger than that of the Ag-CuO or 1Pd:3Ag-CuO. This effect was especially severe at low CuO contents, displaying non-wetting (contact angle > 90°) when the CuO content was less than 10 mol %. Microstructural analysis revealed that this decrease in wettability also caused significant interfacial porosity in the high Pd content (1Pd:1Ag) braze.

ACKNOWLEDGEMENTS

The authors would like to thank Nat Saenz, Shelly Carlson, and Jim Coleman for their assistance in polishing a portion of the wetting samples and conducting the metallographic and SEM analysis work. This work was supported by the U.S. Department of Energy, Office of Fossil Energy, Advanced Research and Technology Development Program. J. Darsell would like to thank his thesis advisors at the School of Mechanical and Materials Engineering at Washington State University, Pullman WA, Amit Bandyapodyay and Susmita Bose. The Pacific Northwest National Laboratory is operated by Battelle Memorial Institute for the United States Department of Energy (U.S. DOE) under Contract DE-AC06-76RLO 1830.

REFERENCES

[1]A.M. Meier, P. R. Chidambaram, G.R. Edwards, "A Comparison of the Wetability of Copper-Copper Oxide and Silver-Copper Oxide on Polycrystalline Alumina," *Journal of Materials Science*, **30** [19] 4781-6 (1995).

[2]C.C. Shüler, A. Stuck, N. Beck, H. Keser, and U. Täck, "Direct Silver Bonding – An Alternative for Substrates in Power Semiconductor Packaging," *Journal of Materials Science: Materials in Electronics*, **11** [3] 389-96 (2000).

[3]K.M. Erskine, MS Thesis, Alfred University, Alfred, NY (1999).

[4]J.Y, Kim, and K.S. Weil, "Development of a Copper Oxide-Silver Braze for Ceramic Joining," Ceramic Transactions: Advances in Joining of Ceramics, Vol. 138, pp.119-132, Edited by C.A. Lewinsohn, M. Singh, and R. Loehman. The American Ceramic Society, (2003).

[5]I. Karayaka and W.T. Thompson, *Bull. Alloy Phase Diagrams*, **9** [3] 237-243 (1988).

[6]Smithell's Metals Reference Book, 6th edition, E. A. Brandes, ed., Butterworths London (1983).

[7]H. Nishiura, R.O. Suzuki, K. Ono, L.J. Gauckler, "Experimental Phase Diagram in the Ag-Cu_2O-Cuo System," *Journal of the American Ceramic Society*, **81** [8] 2181-87 (1998).

Author Index

Keyword Index

9 781574 981797